I0482671

NIST Special Publication 800-57

Recommendation for Key Management – Part 1: General (Revision 3)

Elaine Barker, William Barker, William Burr, William Polk, and Miles Smid

C O M P U T E R   S E C U R I T Y

Computer Security Division
Information Technology Laboratory
National Institute of Standards and Technology
Gaithersburg, MD 20899-8930

July 2012

U.S. Department of Commerce

Rebecca Blank, Acting Secretary

National Institute of Standards and Technology

*Patrick D. Gallagher, Under Secretary for
Standards and Technology, and Director*

Abstract

This Recommendation provides cryptographic key management guidance. It consists of three parts. Part 1 provides general guidance and best practices for the management of cryptographic keying material. Part 2 provides guidance on policy and security planning requirements for U.S. government agencies. Finally, Part 3 provides guidance when using the cryptographic features of current systems.

KEY WORDS: archive; assurances; authentication; authorization; availability; backup; compromise; confidentiality; cryptanalysis; cryptographic key; cryptographic module; digital signature; hash function; key agreement; key management; key management policy; key recovery; key transport; originator-usage period; private key; public key; recipient-usage period; secret key; split knowledge; trust anchor.

Acknowledgements

The National Institute of Standards and Technology (NIST) gratefully acknowledges and appreciates contributions by Lydia Zieglar from the National Security Agency concerning the many security issues associated with this Recommendation. NIST also thanks the many contributions by the public and private sectors whose thoughtful and constructive comments improved the quality and usefulness of this publication.

Authority

This publication has been developed by the National Institute of Standards and Technology (NIST) in furtherance of its statutory responsibilities under the Federal Information Security Management Act (FISMA) of 2002, Public Law 107-347.

NIST is responsible for developing standards and guidelines, including minimum requirements, for providing adequate information security for all agency operations and assets, but such standards and guidelines shall not apply to national security systems.

This Recommendation has been prepared for use by Federal agencies. It may be used by nongovernmental organizations on a voluntary basis and is not subject to copyright. (Attribution would be appreciated by NIST.)

Nothing in this document should be taken to contradict standards and guidelines made mandatory and binding on Federal agencies by the Secretary of Commerce under statutory authority. Nor should these guidelines be interpreted as altering or superseding the existing authorities of the Secretary of Commerce, Director of the OMB, or any other Federal official.

Conformance testing for implementations of this Recommendation will be conducted within the framework of the Cryptographic Algorithm Validation Program (CAVP) and the Cryptographic Module Validation Program (CMVP). The requirements of this Recommendation are indicated by the word "**shall**." Some of these requirements may be out-of-scope for CMVP or CAVP validation testing, and thus are the responsibility of entities using, implementing, installing or configuring applications that incorporate this Recommendation.

Overview

The proper management of cryptographic keys is essential to the effective use of cryptography for security. Keys are analogous to the combination of a safe. If a safe combination is known to an adversary, the strongest safe provides no security against penetration. Similarly, poor key management may easily compromise strong algorithms. Ultimately, the security of information protected by cryptography directly depends on the strength of the keys, the effectiveness of mechanisms and protocols associated with keys, and the protection afforded to the keys. All keys need to be protected against modification, and secret and private keys need to be protected against unauthorized disclosure. Key management provides the foundation for the secure generation, storage, distribution, use and destruction of keys.

Users and developers are presented with many choices in their use of cryptographic mechanisms. Inappropriate choices may result in an illusion of security, but little or no real security for the protocol or application. This Recommendation (i.e., SP 800-57) provides background information and establishes frameworks to support appropriate decisions when selecting and using cryptographic mechanisms.

This Recommendation does not address implementation details for cryptographic modules that may be used to achieve the security requirements identified. These details are addressed in [FIPS140], the associated implementation guidance and the derived test requirements (available at http://csrc.nist.gov/cryptval/).

This Recommendation is written for several different audiences and is divided into three parts.

Part 1, *General*, contains basic key management guidance. It is intended to advise developers and system administrators on the "best practices" associated with key management. Cryptographic module developers may benefit from this general guidance by obtaining a greater understanding of the key management features that are required to support specific, intended ranges of applications. Protocol developers may identify key management characteristics associated with specific suites of algorithms and gain a greater understanding of the security services provided by those algorithms. System administrators may use this document to determine which configuration settings are most appropriate for their information. Part 1 of the Recommendation:

1. Defines the security services that may be provided and key types that may be employed in using cryptographic mechanisms.

2. Provides background information regarding the cryptographic algorithms that use cryptographic keying material.

3. Classifies the different types of keys and other cryptographic information according to their functions, specifies the protection that each type of information requires and identifies methods for providing this protection.

4. Identifies the states in which a cryptographic key may exist during its lifetime.

5. Identifies the multitude of functions involved in key management.

6. Discusses a variety of key management issues related to the keying material. Topics discussed include key usage, cryptoperiod length, domain-parameter validation, public-key validation, accountability, audit, key management system survivability, and guidance for cryptographic algorithm and key size selection.

Part 2, *General Organization and Management Requirements*, is intended primarily to address the needs of system owners and managers. It provides a framework and general guidance to support establishing cryptographic key management within an organization and a basis for satisfying key management aspects of statutory and policy security planning requirements for Federal government organizations.

Part 3, *Implementation-Specific Key Management Guidance*, is intended to address the key management issues associated with currently available implementations.

Table of Contents

Tables

Figures

RECOMMENDATION FOR KEY MANAGEMENT
Part 1: General

1   INTRODUCTION

Cryptographic mechanisms are one of the strongest ways to provide security services for electronic applications and protocols and for data storage. The National Institute of Standards and Technology (NIST) publishes Federal Information Processing Standards (FIPS) and NIST Recommendations (which are published as Special Publications) that specify cryptographic techniques for protecting sensitive, unclassified information.

Since NIST published the Data Encryption Standard (DES) in 1977, the suite of **approved** standardized algorithms has been growing. New classes of algorithms have been added, such as secure hash functions and asymmetric key algorithms for digital signatures. The suite of algorithms now provides different levels of cryptographic strength through a variety of key sizes. The algorithms may be combined in many ways to support increasingly complex protocols and applications. This NIST Recommendation applies to U.S. government agencies using cryptography for the protection of their sensitive, unclassified information. This Recommendation may also be followed, on a voluntary basis, by other organizations that want to implement sound security principles in their computer systems.

The proper management of cryptographic keys is essential to the effective use of cryptography for security. Keys are analogous to the combination of a safe. If the combination is known by an adversary, the strongest safe provides no security against penetration. Similarly, poor key management may easily compromise strong algorithms. Ultimately, the security of information protected by cryptography directly depends on the strength of the keys, the effectiveness of mechanisms and protocols associated with the keys, and the protection afforded the keys. Cryptography can be rendered ineffective by the use of weak products, inappropriate algorithm pairing, poor physical security, and the use of weak protocols.

All keys need to be protected against unauthorized substitution and modification. Secret and private keys need to be protected against unauthorized disclosure. Key management provides the foundation for the secure generation, storage, distribution, and destruction of keys.

1.1   Goal/Purpose

Users and developers are presented with many new choices in their use of cryptographic mechanisms. Inappropriate choices may result in an illusion of security, but little or no real security for the protocol or application. Basic key management guidance is provided in [SP800-21]. This Recommendation (i.e., SP 800-57) expands on that guidance, provides background information and establishes frameworks to support appropriate decisions when selecting and using cryptographic mechanisms.

1.2   Audience

The audiences for this *Recommendation for Key Management* include system or application owners and managers, cryptographic module developers, protocol developers, and system administrators. The Recommendation has been provided in three parts. The different parts into which the Recommendation has been divided have been tailored to specific audiences.

Part 1 of this Recommendation provides general key management guidance that is intended to be useful to both system developers and system administrators. Cryptographic module developers may benefit from this general guidance through a greater understanding of the key management features that are required to support specific intended ranges of applications. Protocol developers may identify key management characteristics associated with specific suites of algorithms and gain a greater understanding of the security services provided by those algorithms. System administrators may use this Recommendation to determine which configuration settings are most appropriate for their information.

Part 2 of this Recommendation is tailored for system or application owners for use in identifying appropriate organizational key management infrastructures, establishing organizational key management policies, and specifying organizational key-management practices and plans.

Part 3 of this Recommendation addresses the key management issues associated with currently available cryptographic mechanisms and is intended to provide guidance to system installers, system administrators and end users of existing key management infrastructures, protocols, and other applications, as well as the people making purchasing decisions for new systems using currently available technology.

Though some background information and rationale are provided for context and to support the recommendations, this document assumes that the reader has a basic understanding of cryptography. For background material, readers may look to a variety of NIST and commercial publications. [SP800-21] includes a brief introduction to cryptography. [SP800-32] provides an introduction to a public-key infrastructure. A mathematical review of cryptography and cryptographic algorithms is found in [HAC] and [AC].

1.3   Scope

This Recommendation encompasses cryptographic algorithms, infrastructures, protocols, and applications, and the management thereof. All cryptographic algorithms currently **approved** by NIST for the protection of unclassified but sensitive information are in scope.

This Recommendation focuses on issues involving the management of cryptographic keys: their generation, use, and eventual destruction. Related topics, such as algorithm selection and appropriate key size, cryptographic policy, and cryptographic module selection, are also included in this Recommendation. Some of the topics noted above are addressed in other NIST standards and guidance. This Recommendation supplements more-focused standards and guidelines.

This Recommendation does not address the implementation details for cryptographic modules that may be used to achieve the security requirements identified. These details are addressed in [SP800-21], [FIPS140], the FIPS 140 implementation guidance and the derived test requirements (available at http://csrc.nist.gov/cryptval/).

This Recommendation also does not address the requirements or procedures for operating an archive, other than discussing the types of keying material that are appropriate to include in an archive and the protection to be provided to the archived keying material.

This Recommendation often uses "requirement" terms; these terms have the following meaning in this document:

1. **shall**: This term is used to indicate a requirement of a Federal Information Processing Standard (FIPS) or a requirement that must be fulfilled to claim conformance to this Recommendation. Note that **shall** may be coupled with **not** to become **shall not**.

2. **should**: This term is used to indicate an important recommendation. Ignoring the recommendation could result in undesirable results. Note that **should** may be coupled with **not** to become **should not**.

1.4   Purpose of FIPS and NIST Recommendations

FIPS security standards and NIST Recommendations are valuable because:

1. They establish an acceptable minimal level of security for U.S. government systems. Systems that implement these Standards and Recommendations offer a consistent level of security **approved** for sensitive, unclassified government data.

2. They often establish some level of interoperability between different systems that implement the Standard or Recommendation. For example, two products that both implement the Advanced Encryption Standard (AES) cryptographic algorithm have the potential to interoperate, provided that the other functions of the product are compatible.

3. They often provide for scalability, because the U.S. government requires products and techniques that can be effectively applied in large numbers.

4. They are scrutinized by the U.S. government to ensure that they provide an adequate level of security. This review is performed by U.S. government experts, in addition to the reviews performed by the public.

5. NIST-**approved** cryptographic techniques are periodically re-assessed for their continued effectiveness. If any technique is found to be inadequate for the continued protection of government information, the Standard or Recommendation is revised or discontinued.

6. Several of the FIPS and NIST Recommendations (e.g., AES, TDEA, SHA-1, and DSA) have required conformance tests. These tests are performed by accredited laboratories on vendor products that claim conformance to the Standards. Vendors are permitted to modify non-conforming products so that they meet all applicable requirements. Users of validated products can have a high degree of confidence that validated products conform to the Standards and Recommendations.

Since 1977, NIST has developed a cryptographic "toolkit" of FIPS security Standards and NIST Recommendations that form a basis for the implementation of **approved** cryptography. This Recommendation references many of those Standards and Recommendations, and provides guidance on how they may be properly used to protect sensitive information.

1.5   Content and Organization

Part 1, *General Guidance*, contains basic key management guidance. It is intended to advise developers and system administrators on the "best practices" associated with key management.

1. Section 1, *Introduction*, establishes the purpose, scope and intended audience of the *Recommendation for Key Management*

2. Section 2, *Glossary of Terms and Acronyms*, provides definitions of terms and acronyms used in this part of the *Recommendation for Key Management*. The reader should be

aware that the terms used in this Recommendation might be defined differently in other documents.

3. Section 3, *Security Services*, defines the security services that may be provided using cryptographic mechanisms.

4. Section 4, *Cryptographic Algorithms*, provides background information regarding the cryptographic algorithms that use cryptographic keying material.

5. Section 5, *General Key Management Guidance,* classifies the different types of keys and other cryptographic information according to their uses, discusses cryptoperiods and recommends appropriate cryptoperiods for each key type, provides recommendations and requirements for other keying material, introduces assurance of domain-parameter and public-key validity, discusses the implications of the compromise of keying material, and provides guidance on cryptographic algorithm strength selection implementation and replacement.

6. Section 6, *Protection Requirements for Cryptographic Information*, specifies the protection that each type of information requires and identifies methods for providing this protection. These protection requirements are of particular interest to cryptographic module vendors and application implementers.

7. Section 7, *Key State and Transitions*, identifies the states in which a cryptographic key may exist during its lifetime.

8. Section 8, *Key Management Phases and Functions*, identifies four phases and a multitude of functions involved in key management. This section is of particular interest to cryptographic module vendors and developers of cryptographic infrastructure services.

9. Section 9, *Accountability, Audit, and Survivability*, discusses three control principles that are used to protect the keying material identified in Section 5.1.

10. Section 10, *Key Management Specifications for Cryptographic Devices or Applications,* specifies the content and requirements for key management specifications. Topics covered include the communications environment, component requirements, keying material storage, access control, accounting, and compromise recovery.

Appendices A and B are provided to supplement the main text where a topic demands a more detailed treatment. Appendix C contains a list of appropriate references, and Appendix D contains a list of changes since the originally published version of this document.

2   Glossary of Terms and Acronyms

The definitions provided below are defined as used in this document. The same terms may be defined differently in other documents.

2.1   Glossary

Access control	Restricts access to resources to only privileged entities.
Accountability	A property that ensures that the actions of an entity may be traced uniquely to that entity.
Algorithm originator-usage period	The period of time during which a specific cryptographic algorithm may be used by originators to apply protection to data.
Algorithm security lifetime	The estimated time period during which data protected by a specific cryptographic algorithm remains secure.
Approved	FIPS-**approved** and/or NIST-recommended. An algorithm or technique that is either 1) specified in a FIPS or NIST Recommendation, or 2) specified elsewhere and adopted by reference in a FIPS or NIST Recommendation.
Archive	To place information into long-term storage. Also, see Key management archive.
Association	A relationship for a particular purpose. For example, a key is associated with the application or process for which it will be used.
Assurance of (private key) possession	Confidence that an entity possesses a private key and its associated keying material.
Assurance of validity	Confidence that a public key or domain parameter is arithmetically correct.
Asymmetric key algorithm	See Public-key cryptographic algorithm.
Attribute	Information associated with a key that is not used in cryptographic algorithms, but is required to implement applications and applications protocols.
Authentication	A process that establishes the source of information, provides assurance of an entity's identity or provides assurance of the integrity of communications sessions, messages, documents or stored data.
Authentication code	A cryptographic checksum based on an **approved** security function (also known as a Message Authentication Code).
Authorization	Access privileges that are granted to an entity; conveying an "official" sanction to perform a security function or activity.
Availability	Timely, reliable access to information by authorized entities.

Backup	A copy of information to facilitate recovery during the cryptoperiod of the key, if necessary.
Certificate	See public-key certificate.
Certification authority	The entity in a Public Key Infrastructure (PKI) that is responsible for issuing certificates and exacting compliance to a PKI policy.
Ciphertext	Data in its encrypted form.
Collision	Two or more distinct inputs produce the same output. Also see hash function.
Compromise	The unauthorized disclosure, modification, substitution or use of sensitive data (e.g., keying material and other security-related information).
Confidentiality	The property that sensitive information is not disclosed to unauthorized entities.
Contingency plan	A plan that is maintained for disaster response, backup operations, and post-disaster recovery to ensure the availability of critical resources and to facilitate the continuity of operations in an emergency situation.
Contingency planning	The development of a contingency plan.
Cryptanalysis	1. Operations performed in defeating cryptographic protection without an initial knowledge of the key employed in providing the protection. 2. The study of mathematical techniques for attempting to defeat cryptographic techniques and information system security. This includes the process of looking for errors or weaknesses in the implementation of an algorithm or in the algorithm itself.
Cryptographic algorithm	A well-defined computational procedure that takes variable inputs, including a cryptographic key, and produces an output.
Cryptographic boundary	An explicitly defined continuous perimeter that establishes the physical bounds of a cryptographic module and contains all hardware, software, and/or firmware components of a cryptographic module.
Cryptographic hash function	See Hash function.

Cryptographic key (key)	A parameter used in conjunction with a cryptographic algorithm that determines its operation in such a way that an entity with knowledge of the key can reproduce or reverse the operation, while an entity without knowledge of the key cannot. Examples include: 1. The transformation of plaintext data into ciphertext data, 2. The transformation of ciphertext data into plaintext data, 3. The computation of a digital signature from data, 4. The verification of a digital signature, 5. The computation of an authentication code from data, 6. The verification of an authentication code from data and a received authentication code, 7. The computation of a shared secret that is used to derive keying material.
Cryptographic key component (key component)	One of at least two parameters that have the same security properties (e.g., randomness) as a cryptographic key; parameters are combined in an **approved** security function to form a plaintext cryptographic key before use.
Cryptographic module	The set of hardware, software, and/or firmware that implements **approved** security functions (including cryptographic algorithms and key generation) and is contained within the cryptographic boundary.
Cryptoperiod	The time span during which a specific key is authorized for use or in which the keys for a given system or application may remain in effect.
Data integrity	A property whereby data has not been altered in an unauthorized manner since it was created, transmitted or stored. In this Recommendation, the statement that a cryptographic algorithm "provides data integrity" means that the algorithm is used to detect unauthorized alterations.
Decryption	The process of changing ciphertext into plaintext using a cryptographic algorithm and key.
Deterministic random bit generator (DRBG)	An algorithm that produces a sequence of bits that are uniquely determined from an initial value called a seed. The output of the DRBG "appears" to be random, i.e., the output is statistically indistinguishable from random values. A cryptographic DRBG has the additional property that the output is unpredictable, given that the seed is not known. A DRBG is sometimes also called a Pseudo Random Number Generator (PRNG) or a deterministic random number generator.

Digital signature	The result of a cryptographic transformation of data that, when properly implemented with a supporting infrastructure and policy, provides the services of: 1. Origin authentication, 2. Data integrity, and 3. Signer non-repudiation.
Distribution	See Key distribution.
Domain parameter	A parameter used in conjunction with some public-key algorithms to generate key pairs, to create digital signatures, or to establish keying material.
Encrypted key	A cryptographic key that has been encrypted using an **approved** security function with a key-encrypting key in order to disguise the value of the underlying plaintext key.
Encryption	The process of changing plaintext into ciphertext using a cryptographic algorithm and key.
Entity	An individual (person), organization, device or process.
Ephemeral key	A cryptographic key that is generated for each execution of a key-establishment process and that meets other requirements of the key type (e.g., unique to each message or session). In some cases, ephemeral keys are used more than once within a single session (e.g., broadcast applications) where the sender generates only one ephemeral key pair per message, and the private key is combined separately with each recipient's public key.
Hash-based message authentication code (HMAC)	A message authentication code that uses an **approved** keyed-hash function (i.e., FIPS 198).
Hash function	A function that maps a bit string of arbitrary length to a fixed-length bit string. **Approved** hash functions satisfy the following properties: 1. (One-way) It is computationally infeasible to find any input that maps to any pre-specified output, and 2. (Collision resistant) It is computationally infeasible to find any two distinct inputs that map to the same output.
Hash value	The result of applying a hash function to information.
Identifier	A bit string that is associated with a person, device or organization. It may be an identifying name, or may be something more abstract (for example, a string consisting of an IP address and timestamp), depending on the application.
Identity	The distinguishing character or personality of an entity.

Initialization vector (IV)	A vector used in defining the starting point of a cryptographic process.
Integrity (also, Assurance of integrity)	See Data integrity.
Key	See Cryptographic key.
Key agreement	A key-establishment procedure where resultant keying material is a function of information contributed by two or more participants, so that no party can predetermine the value of the keying material independently of the other party's contribution.
Key attribute	See Attribute
Key component	See Cryptographic key component.
Key confirmation	A procedure to provide assurance to one party that another party actually possesses the same keying material and/or shared secret.
Key de-registration	A function in the lifecycle of keying material; the marking of all keying material records and associations to indicate that the key is no longer in use.
Key derivation	A function in the lifecycle of keying material; the process by which one or more keys are derived from either a pre-shared key, or a shared secret and other information.
Key-derivation function	A function that, with the input of a cryptographic key or shared secret, and possibly other data, generates a binary string, called keying material.
Key-derivation key	A key used with a key-derivation function or method to derive additional keys. Also called a master key.
Key destruction	To remove all traces of keying material so that it cannot be recovered by either physical or electronic means.
Key distribution	The transport of a key and other keying material from an entity that either owns or generates the key to another entity that is intended to use the key.
Key-encrypting key	A cryptographic key that is used for the encryption or decryption of other keys.
Key establishment	A function in the lifecycle of keying material; the process by which cryptographic keys are securely established among cryptographic modules using manual transport methods (e.g., key loaders), automated methods (e.g., key-transport and/or key-agreement protocols), or a combination of automated and manual methods (consists of key transport plus key agreement).
Key length	Used interchangeably with "Key size".

Key management	The activities involving the handling of cryptographic keys and other related security parameters (e.g., passwords) during the entire lifecycle of the keys, including their generation, storage, establishment, entry and output, use and destruction.
Key management archive	A function in the lifecycle of keying material; a repository for the long-term storage of keying material.
Key Management Policy	A high-level statement of organizational key management policies that identifies a high-level structure, responsibilities, governing Standards and Recommendations, organizational dependencies and other relationships, and security policies.
Key Management Practices Statement	A document or set of documentation that describes in detail the organizational structure, responsible roles, and organization rules for the functions identified in the Key Management Policy.
Key pair	A public key and its corresponding private key; a key pair is used with a public-key algorithm.
Key recovery	A function in the lifecycle of keying material; mechanisms and processes that allow authorized entities to retrieve or reconstruct keying material from key backup or archive.
Key registration	A function in the lifecycle of keying material; the process of officially recording the keying material by a registration authority.
Key revocation	A function in the lifecycle of keying material; a process whereby a notice is made available to affected entities that keying material **should** be removed from operational use prior to the end of the established cryptoperiod of that keying material.
Key size	The length of a key in bits; used interchangeably with "Key length".
Key transport	A key-establishment procedure whereby one party (the sender) selects and encrypts the keying material and then distributes the material to another party (the receiver). When used in conjunction with a public-key (asymmetric) algorithm, the keying material is encrypted using the public key of the receiver and subsequently decrypted using the private key of the receiver. When used in conjunction with a symmetric algorithm, the keying material is encrypted with a key-encrypting key shared by the two parties.
Key update	A function performed on a cryptographic key in order to compute a new, but related, key.
Key-usage period	For a symmetric key, either the originator-usage period or the recipient-usage period.
Key wrapping	A method of encrypting keys (along with associated integrity information) that provides both confidentiality and integrity protection using a symmetric key.

Key-wrapping key	A symmetric key-encrypting key.
Keying material	The data (e.g., keys and IVs) necessary to establish and maintain cryptographic keying relationships.
Manual key transport	A non-automated means of transporting cryptographic keys by physically moving a device, document or person containing or possessing the key or key component.
Master key	See Key-derivation key.
Message authentication code (MAC)	A cryptographic checksum on data that uses a symmetric key to detect both accidental and intentional modifications of data.
Metadata	Information used to describe specific characteristics, constraints, acceptable uses and parameters of another data item (e.g., a cryptographic key).
Non-repudiation	A service that is used to provide assurance of the integrity and origin of data in such a way that the integrity and origin can be verified by a third party as having originated from a specific entity in possession of the private key of the claimed signatory.
Operational phase (Operational use)	A phase in the lifecycle of keying material whereby keying material is used for standard cryptographic purposes.
Operational storage	A function in the lifecycle of keying material; the normal storage of operational keying material during its cryptoperiod.
Owner	For a static key pair, the entity that is associated with the public key and authorized to use the private key. For an ephemeral key pair, the owner is the entity that generated the public/private key pair. For a symmetric key, any entity that is authorized to use the key.
Originator-usage period	The period of time in the cryptoperiod of a symmetric key during which cryptographic protection may be applied to data.
Password	A string of characters (letters, numbers and other symbols) that are used to authenticate an identity, to verify access authorization or to derive cryptographic keys.
Period of protection	The period of time during which the integrity and/or confidentiality of a key needs to be maintained.
Plaintext	Intelligible data that has meaning and can be understood without the application of decryption.

Private key	A cryptographic key, used with a public-key cryptographic algorithm, which is uniquely associated with an entity and is not made public. In an asymmetric (public) cryptosystem, the private key is associated with a public key. Depending on the algorithm, the private key may be used, for example, to: 1. Compute the corresponding public key, 2. Compute a digital signature that may be verified by the corresponding public key, 3. Decrypt keys that were encrypted by the corresponding public key, or 4. Compute a shared secret during a key-agreement transaction.
Proof of possession (POP)	A verification process whereby assurance is obtained that the owner of a key pair actually has the private key associated with the public key.
Pseudorandom number generator (PRNG)	See Deterministic random bit generator (DRBG).
Public key	A cryptographic key, used with a public-key cryptographic algorithm, that is uniquely associated with an entity and that may be made public. In an asymmetric (public) cryptosystem, the public key is associated with a private key. The public key may be known by anyone and, depending on the algorithm, may be used, for example, to: 1. Verify a digital signature that is signed by the corresponding private key, 2. Encrypt keys that can be decrypted using the corresponding private key, or 3. Compute a shared secret during a key-agreement transaction.
Public-key certificate	A set of data that uniquely identifies an entity, contains the entity's public key and possibly other information, and is digitally signed by a trusted party, thereby binding the public key to the entity. Additional information in the certificate could specify how the key is used and its cryptoperiod.
Public-key (asymmetric) cryptographic algorithm	A cryptographic algorithm that uses two related keys: a public key and a private key. The two keys have the property that determining the private key from the public key is computationally infeasible.
Public Key Infrastructure (PKI)	A framework that is established to issue, maintain and revoke public key certificates.
Random bit generator (RBG)	A device or algorithm that outputs a sequence of bits that appear to be statistically independent and unbiased. Also, see Random number generator.

Random number generator (RNG)	A process used to generate an unpredictable series of numbers. Also, referred to as a Random bit generator (RBG).
Recipient-usage period	The period of time during the cryptoperiod of a symmetric key during which the protected information is processed.
Registration authority	A trusted entity that establishes and vouches for the identity of a user.
Retention period	The minimum amount of time that a key or other cryptographically related information should be retained in the archive.
RNG seed	A seed that is used to initialize a deterministic random bit generator. Also called an RBG seed.
Secret key	A cryptographic key that is used with a secret-key (symmetric) cryptographic algorithm that is uniquely associated with one or more entities and is not made public. The use of the term "secret" in this context does not imply a classification level, but rather implies the need to protect the key from disclosure.
Secure communication protocol	A communication protocol that provides the appropriate confidentiality, authentication and content-integrity protection.
Security domain	A system or subsystem that is under the authority of a single trusted authority. Security domains may be organized (e.g., hierarchically) to form larger domains.
Security life of data	The time period during which the security of the data needs to be protected (e.g., its confidentiality, integrity or availability).
Security services	Mechanisms used to provide confidentiality, data integrity, authentication or non-repudiation of information.
Security strength (Also "bits of security")	A number associated with the amount of work (that is, the number of operations) that is required to break a cryptographic algorithm or system. In this Recommendation, the security strength is specified in bits and is a specific value from the set {80, 112, 128, 192, 256}
Seed	A secret value that is used to initialize a process (e.g., a deterministic random bit generator). Also see RNG seed.
Self-signed certificate	A public-key certificate whose digital signature may be verified by the public key contained within the certificate. The signature on a self-signed certificate protects the integrity of the data, but does not guarantee the authenticity of the information. The trust of self-signed certificates is based on the secure procedures used to distribute them.
Shall	This term is used to indicate a requirement of a Federal Information Processing Standard (FIPS) or a requirement that must be fulfilled to claim conformance to this Recommendation. Note that **shall** may be coupled with **not** to become **shall not**.

Shared secret	A secret value that has been computed using a key-agreement scheme and is used as input to a key-derivation function/method.
Should	This term is used to indicate a very important recommendation. Ignoring the recommendation could result in undesirable results. Note that **should** may be coupled with **not** to become **should not**.
Signature generation	The use of a digital signature algorithm and a private key to generate a digital signature on data.
Signature verification	The use of a digital signature algorithm and a public key to verify a digital signature on data.
Split knowledge	A process by which a cryptographic key is split into n multiple key components, individually providing no knowledge of the original key, which can be subsequently combined to recreate the original cryptographic key. If knowledge of k (where k is less than or equal to n) components is required to construct the original key, then knowledge of any k-1 key components provides no information about the original key other than, possibly, its length.
	Note that in this document, split knowledge is not intended to cover key shares, such as those used in threshold or multi-party signatures.
Static key	A key that is intended for use for a relatively long period of time and is typically intended for use in many instances of a cryptographic key-establishment scheme. Contrast with an ephemeral key.
Symmetric key	A single cryptographic key that is used with a secret (symmetric) key algorithm.
Symmetric-key algorithm	A cryptographic algorithm that uses the same secret key for an operation and its complement (e.g., encryption and decryption).
System initialization	A function in the lifecycle of keying material; setting up and configuring a system for secure operation.
Trust anchor	A public key and the name of a certification authority that is used to validate the first certificate in a sequence of certificates. The trust anchor's public key is used to verify the signature on a certificate issued by a trust-anchor certification authority. The security of the validation process depends upon the authenticity and integrity of the trust anchor. Trust anchors are often distributed as self-signed certificates.
Unauthorized disclosure	An event involving the exposure of information to entities not authorized access to the information.
User	See Entity.
User initialization	A function in the lifecycle of keying material; the process whereby a user initializes its cryptographic application (e.g., installing and initializing software and hardware).

User registration	A function in the lifecycle of keying material; a process whereby an entity becomes a member of a security domain.
Work	The expected time to break a cipher with a given resource. For example, 12 MIPS years would be the amount of work that one computer, with the capability of processing a million instructions per second, could do in 12 years. The same amount of work could be done by 12 such computers in one year, assuming that the algorithm being executed can be sufficiently parallelized.
X.509 certificate	The X.509 public-key certificate or the X.509 attribute certificate, as defined by the ISO/ITU-T X.509 standard. Most commonly (including in this document), an X.509 certificate refers to the X.509 public-key certificate.
X.509 public-key certificate	A digital certificate containing a public key for entity and a name for the entity, together with some other information that is rendered un-forgeable by the digital signature of the certification authority that issued the certificate, encoded in the format defined in the ISO/ITU-T X.509 standard.

2.2   Acronyms

The following abbreviations and acronyms are used in this Recommendation:

2TDEA   Two-key Triple Data Encryption Algorithm

3TDEA   Three-key Triple Data Encryption Algorithm

AES   Advanced Encryption Standard specified in [FIPS197].

ANS   American National Standard

ANSI   American National Standards Institute

CA   Certification Authority

CRC   Cyclic Redundancy Check

CRL   Certificate Revocation List

DRBG   Deterministic Random Bit Generator

DSA   Digital Signature Algorithm specified in [FIPS186]

ECC   Elliptic Curve Cryptography

ECDSA   Elliptic Curve Digital Signature Algorithm specified in [ANSX9.62]

FFC   Finite Field Cryptography

FIPS   Federal Information Processing Standard

HMAC   Keyed-Hash Message Authentication Code specified in [FIPS198]

IFC   Integer Factorization Cryptography

IV   Initialization Vector

MAC	Message Authentication Code
NIST	National Institute of Standards and Technology
PKI	Public-Key Infrastructure
POP	Proof of Possession
RA	Registration Authority
RBG	Random Bit Generator
RNG	Random Number Generator
RSA	Rivest, Shamir, Adelman (an algorithm)
TDEA	Triple Data Encryption Algorithm; Triple DEA
TLS	Transport Layer Security

3 Security Services

Cryptography may be used to perform several basic security services: confidentiality, data integrity, authentication, authorization and non-repudiation. These services may also be required to protect cryptographic keying material. In addition, there are other cryptographic and non-cryptographic mechanisms that are used to support these security services. In general, a single cryptographic mechanism may provide more than one service (e.g., the use of digital signatures can provide integrity, authentication and non-repudiation), but not all services.

3.1 Confidentiality

Confidentiality is the property whereby information is not disclosed to unauthorized parties. Secrecy is a term that is often used synonymously with confidentiality. Confidentiality is achieved using encryption to render the information unintelligible except by authorized entities. The information may become intelligible again by using decryption. In order for encryption to provide confidentiality, the cryptographic algorithm and mode of operation must be designed and implemented so that an unauthorized party cannot determine the secret or private keys associated with the encryption or be able to derive the plaintext directly without deriving any keys.

3.2 Data Integrity

Data integrity is a property whereby data has not been altered in an unauthorized manner since it was created, transmitted or stored. Alteration includes the insertion, deletion and substitution of data. Cryptographic mechanisms, such as message authentication codes or digital signatures, can be used to detect (with a high probability) both accidental modifications (e.g., modifications that sometimes occur during noisy transmissions or by hardware memory failures) and deliberate modifications by an adversary. Non-cryptographic mechanisms are also often used to detect accidental modifications, but cannot be relied upon to detect deliberate modifications. A more detailed treatment of this subject is provided in Appendix A.1.

In this Recommendation, the statement that a cryptographic algorithm "provides data integrity" means that the algorithm is used to detect unauthorized alterations.

3.3 Authentication

Authentication is a service that is used to establish the origin and integrity of information. That is, authentication services verify the identity of the user or system that created information (e.g., a transaction or message) or verify that the data has not been modified. This service supports the receiver in security-relevant decisions, such as "Is the sender an authorized user of this system?" or "Is the sender permitted to read sensitive information?" Several cryptographic mechanisms may be used to provide authentication services. Most commonly, authentication is provided by digital signatures or message authentication codes; some key-agreement techniques also provide authentication. When multiple individuals are permitted to share the same authentication information (such as a password or cryptographic key), it is sometimes called role-based authentication. See [FIPS140].

3.4 Authorization

Authorization is concerned with providing an official sanction or permission to perform a security function or activity. Normally, authorization is granted following a process of authentication. A non-cryptographic analog of the interaction between authentication and authorization is the examination of an individual's credentials to establish their identity (authentication); upon proving identity, the individual is then provided with the key or password that will allow access to some resource, such as a locked room (authorization). Authentication can be used to authorize a role, rather than to identify an individual. Once authenticated to a role, an entity is authorized for all the privileges associated with the role.

3.5   Non-repudiation

Non-repudiation is a service that is used to provide assurance of the integrity and origin of data in such a way that the integrity and origin can be verified by a third party. This service prevents an entity from successfully denying involvement in a previous action. Non-repudiation is supported cryptographically by the use of a digital signature that is calculated by a private key known only by the entity that computes the digital signature.

3.6   Support Services

Cryptographic security services often require supporting services. For example, cryptographic services often require the use of key establishment and random number generation services.

3.7   Combining Services

In many applications, a combination of cryptographic services (confidentiality, data integrity, authentication, authorization, and non-repudiation) is desired. Designers of secure systems often begin by considering which security services are needed to protect the information contained within and processed by the system. After these services have been determined, the designer then considers what mechanisms will best provide these services. Not all mechanisms are cryptographic in nature. For example, physical security may be used to protect the confidentiality of certain types of data, and identification badges or biometric identification devices may be used for entity authentication. However, cryptographic mechanisms consisting of algorithms, keys, and other keying material often provide the most cost-effective means of protecting the security of information. This is particularly true in applications where the information would otherwise be exposed to unauthorized entities.

When properly implemented, some cryptographic algorithms provide multiple services. The following examples illustrate this case:

1. A message authentication code (Section 4.2.3) can provide authentication, as well as data integrity if the symmetric keys are unique to each pair of users.

2. A digital signature algorithm (Section 4.2.4) can provide authentication and data integrity, as well as non-repudiation.

3. Certain modes of encryption can provide confidentiality, data integrity, and authentication when properly implemented. These modes **should** be specifically designed to provide these services.

However, it is often the case that different algorithms need to be employed in order to provide all the desired services.

Example:

Consider a system where the secure exchange of information between pairs of Internet entities is needed. Some of the exchanged information requires just integrity protection, while other information requires both integrity and confidentiality protection. It is also a requirement that each entity that participates in an information exchange knows the identity of the other entity.

The designers of this example system decide that a Public Key Infrastructure (PKI) needs to be established and that each entity wishing to communicate securely is required to physically authenticate his or her identity at a Registration Authority (RA). This authentication requires the presentation of proper credentials, such as a driver's license, passport or birth certificate. The authenticated individuals then generate a public static key pair in a smart card that is later used for key agreement. The public static key-agreement key of each net member is transferred from the smart card to the RA, where it is incorporated with the user identifier and other information into a digitally signed message for transmission to a Certification Authority (CA). The CA then composes the user's public-key certificate by signing the public key of the user and the user's identifier, along with other information. This certificate is returned to the public-key owner so that it may be used in conjunction with the private key (under the sole control of the owner) for entity-authentication and key-agreement purposes.

In this example, any two entities wishing to communicate may exchange public-key certificates containing public keys that are checked by verifying the CA's signature on the certificate (using the CA's public key). The public static key-agreement key of each of the two entities and each entity's own private static key-agreement key are then used in a key-agreement scheme to produce a shared secret that is known by the two entities. The shared secret may then be used to derive one or more shared symmetric keys. If the mode of the symmetric-encryption algorithm is designed to support all the desired services, then only one shared key is necessary. Otherwise, multiple shared keys and algorithms are used, e.g., one of the shared keys is used to encrypt for confidentiality, while another key is used for integrity and authentication. The receiver of the data protected by the key(s) has assurance that the data came from the other entity indicated by the public-key certificate, that the data remains confidential, and that the integrity of the data is preserved.

Alternatively, if confidentiality is not required, integrity protection, entity authentication, and non-repudiation can be attained by establishing a digital-signature key pair and corresponding certificate for each entity. The private signature key of the sender is used to sign the data, and the sender's public signature-verification key is used by the receiver to verify the signature. In this case, a single algorithm provides all three services.

The above example provides a basic sketch of how cryptographic algorithms may be used to support multiple security services. However, it can be easily seen that the security of such a system depends on many factors, including:

a. The strength of the entity's credentials (e.g., driver's license, passport or birth certificate) and authentication mechanism,

b. The strength of the cryptographic algorithms used,

c. The degree of trust placed in the RA and the CA,

d. The strength of the key-establishment protocols, and

e. The care taken by the users in protecting their keys from unauthorized use.

Therefore, the design of a security system that provides the desired security services by making use of cryptographic algorithms and sound key management techniques requires a high degree of skill and expertise.

4   Cryptographic Algorithms

FIPS-**approved** or NIST-recommended cryptographic algorithms **shall** be used whenever cryptographic services are required. These **approved** algorithms have received an intensive security analysis prior to their approval and continue to be examined to determine that the algorithms provide adequate security. Most cryptographic algorithms require cryptographic keys or other keying material. In some cases, an algorithm may be strengthened by the use of larger keys. This Recommendation advises the users of cryptographic mechanisms on the appropriate choices of algorithms and key sizes.

This section describes the **approved** cryptographic algorithms that provide security services, such as confidentiality, data integrity, authentication, authorization, non-repudiation.

4.1   Classes of Cryptographic Algorithms

There are three basic classes of **approved** cryptographic algorithms: hash functions, symmetric-key algorithms and asymmetric-key algorithms. The classes are defined by the number of cryptographic keys that are used in conjunction with the algorithm.

Cryptographic hash functions do not require keys. Hash functions generate a relatively small digest (hash value) from a (possibly) large input in a way that is fundamentally difficult to reverse (i.e., it is hard to find an input that will produce a given output). Hash functions are used as building blocks for key management, for example,

1. To provide data authentication and integrity services (Section 4.2.3) – the hash function is used with a key to generate a message authentication code;

2. To compress messages for digital signature generation and verification (Section 4.2.4);

3. To derive keys in key-establishment algorithms (Section 4.2.5); and

4. To generate deterministic random numbers (Section 4.2.7).

Symmetric-key algorithms (sometimes known as secret-key algorithms) transform data in a way that is fundamentally difficult to undo without knowledge of a secret key. The key is "symmetric" because the same key is used for a cryptographic operation and its inverse (e.g., encryption and decryption). Symmetric keys are often known by more than one entity; however, the key **shall not** be disclosed to entities that are not authorized access to the data protected by that algorithm and key. Symmetric key algorithms are used, for example,

1. To provide data confidentiality (Section 4.2.2); the same key is used to encrypt and decrypt data;

2. To provide authentication and integrity services (Section 4.2.3) in the form of Message Authentication Codes (MACs); the same key is used to generate the MAC and to validate it. MACs normally employ either a symmetric key-encryption algorithm or a cryptographic hash function as their cryptographic primitive;

3. As part of the key-establishment process (Section 4.2.5); and

4. To generate deterministic random numbers (Section 4.2.7).

Asymmetric-key algorithms, commonly known as public-key algorithms, use two related keys (i.e., a key pair) to perform their functions: a public key and a private key. The public key may be known by anyone; the private key **should**[1] be under the sole control of the entity that "owns" the key pair. Even though the public and private keys of a key pair are related, knowledge of the public key does not reveal the private key. Asymmetric algorithms are used, for example,

1. To compute digital signatures (Section 4.2.4);

2. To establish cryptographic keying material (Section 4.2.5); and

3. To generate random numbers (Section 4.2.7).

4.2 Cryptographic Algorithm Functionality

Security services are fulfilled using a number of different algorithms. In many cases, the same algorithm may be used to provide multiple services.

4.2.1 Hash Functions

Many algorithms and schemes that provide a security service use a hash function as a component of the algorithm. Hash functions can be found in digital signature algorithms (see [FIPS186]), Keyed-Hash Message Authentication Codes (HMAC) (see [FIPS198]), key-derivation functions/methods (see [SP800-56A], [SP800-56B], [SP800-56C] and [SP800-108]), and random number generators (see [SP800-90A]). **Approved** hash functions are defined in [FIPS180].

A hash function takes an input of arbitrary length and outputs a fixed-length value. Common names for the output of a hash function include hash value, hash, message digest, and digital fingerprint. The maximum number of input and output bits is determined by the design of the hash function. All **approved** hash functions are cryptographic hash functions. With a well-designed cryptographic hash function, it is not feasible to find a message that will produce a given hash value (pre-image resistance), nor is it feasible to find two messages that produce the same hash value (collision resistance).

Several hash functions are **approved** for Federal Government use and are defined in [FIPS180], including SHA-1, SHA-224, SHA-512/224, SHA-256, SHA-512/256, SHA-384 and SHA-512[2]. The size of the hash value produced by SHA-1 is 160 bits; 224 bits for SHA-224 and SHA-512/224; 256 bits for SHA-256 and SHA-512/256; 384 bits for SHA-384, and 512 bits for SHA-512. Algorithm standards need to specify either the appropriate size for the hash function or provide the hash-function selection criteria if the algorithm can be configured to use different hash functions.

4.2.2 Symmetric-Key Algorithms used for Encryption and Decryption

Encryption is used to provide confidentiality for data. The data to be protected is called plaintext when in its original form. Encryption transforms the data into ciphertext. Ciphertext can be transformed back into plaintext using decryption. The **approved** algorithms for encryption/decryption are symmetric key algorithms: AES and TDEA. Each of these algorithms

---

[1] Sometimes a key pair is generated by a party that is trusted by the key owner.

[2] In general the notation SHA-n indicates a hash function specified in [FIPS180] that provides an n-bit hash value. However, SHA-1 indicates a hash function with a 160-bit hash value that was originally specified in FIPS 180-1.

operates on blocks (chunks) of data during an encryption or decryption operation. For this reason, these algorithms are commonly referred to as block cipher algorithms.

4.2.2.1   Advanced Encryption Standard (AES)

The AES algorithm is specified in [FIPS197]. AES encrypts and decrypts data in 128-bit blocks, using 128, 192 or 256-bit keys. The nomenclature for AES for the different key sizes is AES-*x*, where *x* is the key size. All three key sizes are considered adequate for Federal Government applications.

4.2.2.2   Triple DEA (TDEA)

Triple DEA is defined in [SP800-67]. TDEA encrypts and decrypts data in 64-bit blocks, using three 56-bit keys. Federal applications **should** use three distinct keys.

4.2.2.3   Modes of Operation

With a block-cipher encryption operation, the same plaintext block will always encrypt to the same ciphertext block whenever the same key is used. If the multiple blocks in a typical message are encrypted separately, an adversary can easily substitute individual blocks, possibly without detection. Furthermore, certain kinds of data patterns in the plaintext, such as repeated blocks, are apparent in the ciphertext.

Cryptographic modes of operation have been defined to alleviate this problem by combining the basic cryptographic algorithm with variable initialization vectors and some sort of feedback of the information derived from the cryptographic operation. The NIST Recommendation for Block Cipher Modes of Operation [SP800-38A] defines modes of operation for the encryption and decryption of data using block cipher algorithms, such as AES and TDEA. Other modes **approved** for encryption are specified in other parts of [SP800-38]; some of these modes also produce message authentication codes (see Section 4.2.3).

4.2.3   Message Authentication Codes (MACs)

Message Authentication Codes (MACs) provide data authentication and integrity. A MAC is a cryptographic checksum on the data that is used in order to provide assurance that the data has not changed and that the MAC was computed by the expected entity. Although message integrity is often provided using non-cryptographic techniques known as error detection codes, these codes can be altered by an adversary to effect an action to the adversary's benefit. The use of an **approved** cryptographic mechanism, such as a MAC, can alleviate this problem. In addition, the MAC can provide a recipient with assurance that the originator of the data is a key holder (i.e., an entity authorized to have the key). MACs are often used to authenticate the originator to the recipient when only those two parties share the MAC key.

The computation of a MAC requires the use of (1) a secret key that is known only by the party that generates the MAC and by the intended recipient(s) of the MAC, and (2) the data on which the MAC is calculated. The result of the MAC computation is often called a MacTag when transmitted; a MacTag is the full-length or truncated result from the MAC computation. Two types of algorithms for computing a MAC have been **approved**: MAC algorithms that are based on block cipher algorithms, and MAC algorithms that are based on hash functions.

4.2.3.1   MACs Using Block Cipher Algorithms

[SP800-38B] defines a mode to compute a MAC using **approved** block cipher algorithms, such as AES and TDEA. The key and block size used to compute the MAC depend on the algorithm used. If the same block cipher is used for both encryption and MAC computation in two separate cryptographic operations, then the same key **shall not** be used for both the MAC and encryption operations. Note that some modes of operation specified in [SP800-38] perform encryption and message authentication using a single key.

4.2.3.2   MACs Using Hash Functions

[FIPS198] specifies the computation of a MAC using an **approved** hash function. The algorithm requires a single pass through the entire data. A variety of key sizes are allowed for HMAC; the choice of key size depends on the amount of security to be provided to the data and the hash function used. See [SP800-107] for further discussions about HMAC, and Section 5.6 of this Recommendation (i.e., SP 800-57, Part 1) for guidance in the selection of key sizes.

4.2.4   Digital Signature Algorithms

Digital signatures are used to provide authentication, integrity and non-repudiation. Digital signatures are used in conjunction with hash functions and are computed on data of any length (up to a limit that is determined by the hash function). [FIPS186] specifies algorithms that are **approved** for the computation of digital signatures[3]. It defines the Digital Signature Algorithm (DSA) and adopts the RSA algorithm as specified in [ANSX9.31] and [PKCS#1] (version 1.5 and higher), and the ECDSA algorithm as specified in [ANSX9.62].

4.2.4.1   DSA

The Digital Signature Algorithm (DSA) is specified in [FIPS186] for specific key sizes[4]: 1024, 2048, and 3072 bits. The DSA will produce digital signatures of 320, 448, or 512 bits[5]. Older systems (legacy systems) used smaller key sizes. While it may be appropriate to continue to verify and honor signatures created using these smaller key sizes[6], new signatures **shall not** be created using these key sizes.

4.2.4.2   RSA

The RSA algorithm, as specified in [ANSX9.31] and [PKCS#1] (version 1.5 and higher) is adopted for the computation of digital signatures in [FIPS186]. [FIPS186] specifies methods for generating RSA key pairs for several key sizes for [ANSX9.31] and [PKCS#1] implementations. Older systems (legacy systems) used smaller key sizes. While it may be appropriate to continue

---

[3] Two general types of digital signature methods are discussed in literature: digital signatures with appendix, and digital signatures with message recovery. [FIPS186] specifies algorithms for digital signatures with appendix, and is the digital signature method that is discussed in this Recommendation.

[4] For DSA, the key size is considered to be the size of the modulus p. Another value, q, is also important when defining the security afforded by DSA.

[5] The length of the digital signature is twice the size of q (see the previous footnote).

[6] This may be appropriate if it is possible to determine when the digital signature was created (e.g., by the use of a trusted time stamping process). The decision should not depend solely on the cryptography used.

to verify and honor signatures created using these smaller key sizes[7], new signatures **shall not** be created using these key sizes.

4.2.4.3   ECDSA

The Elliptic Curve Digital Signature Algorithm (ECDSA), as specified in [ANSX9.62], is adopted for the computation of digital signatures in [FIPS186]. [ANSX9.62] specifies a minimum key size[8] of 160 bits. ECDSA produces digital signatures that are twice the length of the key size. Recommended elliptic curves are provided in [FIPS186].

4.2.5   Key Establishment Schemes

Automated key-establishment schemes are used to set up keys to be used between communicating entities. Two types of automated key-establishment schemes are defined: key transport and key agreement. **Approved** key-establishment schemes are provided in [SP800-56A] and [SP800-56B].

Key transport is the distribution of a key (and other keying material) from one entity to another entity. The keying material is usually encrypted by the sending entity and decrypted by the receiving entity(ies). If a symmetric algorithm (e.g., AES key wrap) is used to encrypt the keying material to be distributed, the sending and receiving entities need to know the symmetric key-wrapping key (i.e., the key-encrypting key). If a public-key algorithm is used to distribute the keying material, a key pair is used as the key-encrypting key; in this case, the sending entity encrypts the keying material using the receiving entity's public key; the receiving entity decrypts the received keying material using the associated private key.

Key agreement is the participation by both entities (i.e., the sending and receiving entities) in the creation of shared keying material. This may be accomplished using either asymmetric (public-key) or symmetric key techniques. If an asymmetric algorithm is used, each entity has either a static key pair or an ephemeral key pair or both. If a symmetric algorithm is used, each entity shares the same symmetric key-wrapping key.

[SP800-56A] specifies key-establishment schemes that use discrete-logarithm-based public-key algorithms. With the key-establishment schemes specified in [SP800-56A], a party may own an ephemeral key, a static key, or both an ephemeral and a static key. The ephemeral key is used to provide a new secret for each key-establishment transaction, while the static key (if used in a PKI with public-key certificates) provides for the authentication of the owner. [SP800-56A] characterizes each scheme into a class, depending upon how many ephemeral and static keys are used. Each scheme class has its corresponding security properties.

[SP800-56B] provides key-establishment schemes that use integer-factorization-based public-key algorithms. Two of the families of schemes specified in [SP800-56B] provide for key agreement, and the other two families provide for key transport. In these schemes, one party always owns a key pair, and the other party may or may not own a key pair, depending on the scheme. In these schemes, only static keys are used; ephemeral keys are not used.

---

[7] This may be appropriate if it is possible to determine when the digital signature was created (e.g., by the use of a trusted time stamping process). The decision should not depend solely on the cryptography used.

[8] For elliptic curves, the key size is the length f of the order n of the base point G of the chosen elliptic curve.

Cryptographic protocol designers need to understand the security properties of the schemes in order to assure that the desired capabilities are available to the user. In general, schemes where each party uses both an ephemeral and a static key provide more security properties than schemes using fewer keys. However, it may not be practical for both parties to use both static and ephemeral keys in certain applications, and the use of ephemeral keys is not specified for all algorithms (see [SP800-56B]). For example, in email applications, it is desirable to send messages to other parties who are not on-line. In this case the receiver cannot be expected to use an ephemeral key to establish the message-encrypting key.

4.2.5.1 Discrete Log Key Agreement Schemes Using Finite Field Arithmetic

Key agreement schemes based on the intractability of the discrete-logarithm problem and using finite-field arithmetic have been specified in [SP800-56A]. Each scheme provides a different configuration of required key pairs that may be used, depending on the requirements of a communication situation.

4.2.5.2 Discrete Log Key Agreement Schemes Using Elliptic Curve Arithmetic

Key agreement schemes based on the intractability of the discrete-logarithm problem and using elliptic-curve arithmetic have been specified in [SP800-56A]. Each scheme provides a different configuration of required key pairs that may be used, depending on the requirements of a communication situation.

4.2.5.3 RSA Key Establishment

RSA key-establishment schemes based on the integer-factorization problem have been **approved** in [SP800-56B]. Four scheme families are specified, two families for key agreement and two for key transport. Each scheme family has a basic scheme and one or more key confirmation schemes.

4.2.5.4 Key Wrapping

Key wrapping is the encryption of a key by a key-encrypting key using a symmetric algorithm (e.g., an AES key is encrypted by an AES key-encrypting key). Key wrapping provides both confidentiality and integrity to the wrapped material. Several methods for key wrapping have been specified or referenced in [SP800-38F].

4.2.5.5 Key Confirmation

Key confirmation is used by two parties in a key-establishment process to provide assurance that common keying material and/or a shared secret has been established. The assurance may be provided to only one party (unilateral) or it may be provided to both parties (bilateral). The assurance may be provided as part of the key-establishment scheme or it may be provided by some action that takes place outside of the scheme. For example, after a key is established, two parties may confirm to one another that they possess the same secret by demonstrating their ability to encrypt and decrypt data intended for each other.

[SP800-56A] provides for unilateral key confirmation for schemes where one party has a static key-establishment key, and bilateral key confirmation for schemes where both parties have static key-establishment keys. A total of ten key confirmation schemes are provided, seven of which are unilateral, and three of which are bilateral.

[SP800-56B] provides for unilateral key confirmation from the responder, in the case of key agreement, and from the receiver, in the case of key transport. Initiator and bilateral key confirmation are also provided for one family of key confirmation schemes.

4.2.6   Key Establishment Protocols

Key establishment protocols use key-establishment schemes in order to specify the processing necessary to establish a key. However, key-establishment protocols also specify message flow and format. Key-establishment protocols need to be carefully designed to not give secret information to a potential attacker. For example, a protocol that indicates abnormal conditions, such as a data integrity error, may permit an attacker to confirm or reject an assumption regarding secret data. Alternatively, if the time or power required to perform certain computations are based upon the value of the secret or private key in use, then an attacker may be able to deduce the key from observed fluctuations.

Therefore, it is best to design key-establishment protocols so that:

1. The protocols do not provide for an early exit from the protocol upon detection of a single error,

2. The protocols trigger an alarm after a certain reasonable number of detected error conditions, and

3. The key-dependent computations are obscured from the observer in order to prevent or minimize the detection of key-dependent characteristics.

4.2.7   Random Number Generation

Random bit generators (RBGs) and random number generators (RNGs) are required for the generation of keying material (e.g., keys and IVs). RBGs generate sequences of random bits (e.g., 010011); technically, RNGs translate those bits into numbers (e.g., 010011 is translated into the number 19). However, the use of the term "random number generator" (RNG) is commonly used to refer to both concepts, and will be used interchangeably with "RBG" in this document.

Two classes of RBGs are defined: deterministic and non-deterministic. Deterministic Random Bit Generators (DRBGs), sometimes called deterministic random number generators or pseudorandom number generators, use cryptographic algorithms and the associated keying material to generate random bits; Non-deterministic Random Bit Generators (NRBGs), sometimes called true RNGs, produce output that is dependent on some unpredictable physical source that is outside human control.

[SP800-90A] specifies DRBGs that may be used to generate random bits for cryptographic applications (e.g., key or IV generation). A DRBG is initialized with a secret starting value, called a seed. An "attacker" with knowledge of the DRBG output should not be able to determine the seed other than by exhaustive guessing.

5 GENERAL KEY MANAGEMENT GUIDANCE

This section classifies the different types of keys and other cryptographic information according to their uses; discusses cryptoperiods and recommends appropriate cryptoperiods for each key type; provides recommendations and requirements for other keying material; introduces assurance of domain-parameter validity, public-key validity, and private-key possession; discusses the implications of the compromise of keying material; and provides guidance on the selection, implementation, and replacement of cryptographic algorithms and key sizes according to their security strengths.

5. 1 Key Types and Other Information

There are several different types of cryptographic keys, each used for a different purpose. In addition, there is other information that is specifically related to cryptographic algorithms and keys.

5.1.1 Cryptographic Keys

Several different types of keys are defined. The keys are identified according to their classification as public, private or symmetric keys, and as to their use. For public and private key-agreement keys, their status as static or ephemeral keys is also specified. See Table 5 in Section 6.1.1 for the required protections for each type of information.

1. *Private signature key*: Private signature keys are the private keys of asymmetric (public) key pairs that are used by public-key algorithms to generate digital signatures with possible long-term implications. When properly handled, private signature keys can be used to provide source authentication, integrity protection and non-repudiation of messages, documents or stored data.

2. *Public signature-verification key*: A public signature-verification key is the public key of an asymmetric (public) key pair that is used by a public-key algorithm to verify digital signatures that are intended to provide source authentication, integrity protection and non-repudiation of messages, documents or stored data.

3. *Symmetric authentication key*: Symmetric authentication keys are used with symmetric-key algorithms to provide source authentication and assurance of the integrity of communication sessions, messages, documents or stored data.

4. *Private authentication key*: A private authentication key is the private key of an asymmetric (public) key pair that is used with a public-key algorithm to provide assurance of the identity of the originating entity when executing an authentication mechanism as part of an authentication protocol run or when establishing an authenticated communication session.

5. *Public authentication key*: A public authentication key is the public key of an asymmetric (public) key pair that is used with a public-key algorithm to provide assurance of the identity of the originating entity when executing an authentication mechanism as part of an authentication protocol run or when establishing an authenticated communication session.

6. *Symmetric data-encryption key*: These keys are used with symmetric-key algorithms to apply confidentiality protection to information.

7. *Symmetric key-wrapping key*: Symmetric key-wrapping keys are used to encrypt other keys using symmetric-key algorithms. Key-wrapping keys are also known as key-encrypting keys.

8. *Symmetric and asymmetric random number generation keys*: These keys are keys used to generate random numbers.

9. *Symmetric master key*: A symmetric master key is used to derive other symmetric keys (e.g., data-encryption keys, key-wrapping keys, or authentication keys) using symmetric cryptographic methods. The master key is also known as a key-derivation key.

10. *Private key-transport key*: Private key-transport keys are the private keys of asymmetric (public) key pairs that are used to decrypt keys that have been encrypted with the associated public key using a public-key algorithm. Key-transport keys are usually used to establish keys (e.g., key-wrapping keys, data encryption keys or MAC keys) and, optionally, other keying material (e.g., Initialization Vectors).

11. *Public key-transport key*: Public key-transport keys are the public keys of asymmetric (public) key pairs that are used to encrypt keys using a public-key algorithm. These keys are used to establish keys (e.g., key-wrapping keys, data-encryption keys or MAC keys) and, optionally, other keying material (e.g., Initialization Vectors).

12. *Symmetric key-agreement key*: These symmetric keys are used to establish keys (e.g., key-wrapping keys, data-encryption keys, or MAC keys) and, optionally, other keying material (e.g., Initialization Vectors) using a symmetric key-agreement algorithm.

13. *Private static key-agreement key*: Private static key-agreement keys are the private keys of asymmetric (public) key pairs that are used to establish keys (e.g., key-wrapping keys, data-encryption keys, or MAC keys) and, optionally, other keying material (e.g., Initialization Vectors).

14. *Public static key-agreement key*: Public static key-agreement keys are the public keys of asymmetric (public) key pairs that are used to establish keys (e.g., key-wrapping keys, data-encryption keys, or MAC keys) and, optionally, other keying material (e.g., Initialization Vectors).

15. *Private ephemeral key-agreement key*: Private ephemeral key-agreement keys are the private keys of asymmetric (public) key pairs that are used only once[9] to establish one or more keys (e.g., key-wrapping keys, data-encryption keys, or MAC keys) and, optionally, other keying material (e.g., Initialization Vectors).

16. *Public ephemeral key-agreement key*: Public ephemeral key-agreement keys are the public keys of asymmetric key pairs that are used in a single key-establishment

---

[9] In some cases ephemeral keys are used more than once, though within a single "session". For example, when Diffie-Hellman is used in S/MIME CMS, the sender may generate one ephemeral key pair per message, and the private key is combined separately with each recipient's public key.

transaction[10] to establish one or more keys (e.g., key-wrapping keys, data-encryption keys, or MAC keys) and, optionally, other keying material (e.g., Initialization Vectors).

17. *Symmetric authorization key*: Symmetric authorization keys are used to provide privileges to an entity using a symmetric cryptographic method. The authorization key is known by the entity responsible for monitoring and granting access privileges for authorized entities and by the entity seeking access to resources.

18. *Private authorization key*: A private authorization key is the private key of an asymmetric (public) key pair that is used to provide privileges to an entity.

19. *Public authorization key*: A public authorization key is the public key of an asymmetric (public) key pair that is used to verify privileges for an entity that knows the associated private authorization key.

5.1.2   Other Cryptographic or Related Information

Other information used in conjunction with cryptographic algorithms and keys also needs to be protected. See Table 6 in Section 6.1.2 for the required protections for each type of information.

1. *Domain Parameters*: Domain parameters are used in conjunction with some public-key algorithms to generate key pairs, to create digital signatures or to establish keying material.

2. *Initialization Vectors*: Initialization vectors (IVs) are used by several modes of operation for encryption and decryption (see Section 4.2.2.3) and for the computation of MACs using block cipher algorithms (see Section 4.2.3.1)

3. *Shared Secrets:* Shared secrets are generated during a key-establishment process as defined in [SP800-56A] and [SP800-56B]. Shared secrets **shall** be protected and handled in the same manner as cryptographic keys. If a FIPS 140-validated cryptographic module is being used, then the protection of the shared secrets is provided by the cryptographic module.

4. *RNG seeds*: RNG seeds are used in the generation of *deterministic random* numbers (e.g., used to generate keying material that must remain secret or private).

5. *Other public information*: Public information (e.g., a nonce) is often used in the key-establishment process.

6. *Other secret information*: Secret information may be included in the seeding of an RNG or in the establishment of keying material.

7. *Intermediate Results*: The intermediate results of cryptographic operations using secret information must be protected. Intermediate results **shall not** be available for purposes other than as intended.

8. *Key control information*: Information related to the keying material (e.g., the identifier, purpose, or a counter) must be protected to ensure that the associated keying material can be correctly used. The key control information is included in the metadata associated with the key (see Section 6.2.3.1).

---

[10] The public ephemeral key-agreement key of a sender may be retained by the receiver for later use in decrypting a stored (encrypted) message for which the ephemeral key pair was generated.

9. *Random numbers*: The random numbers created by a random number generator **should** be protected when retained. When used directly as keying material, the random numbers **shall** be protected as discussed in Section 6.

10. *Passwords*: A password is used to acquire access to privileges and can be used as a credential in an authentication mechanism. A password can also be used to derive cryptographic keys that are used to protect and access data in storage, as specified in [SP800-132].

11. *Audit information*: Audit information contains a record of key management events.

5.2   Key Usage

In general, a single key **should** be used for only one purpose (e.g., encryption, authentication, key wrapping, random number generation, or digital signatures). There are several reasons for this:

1.   The use of the same key for two different cryptographic processes may weaken the security provided by one or both of the processes.

2.   Limiting the use of a key limits the damage that could be done if the key is compromised.

3.   Some uses of keys interfere with each other. For example, consider a key pair used for both key transport and digital signatures. In this case, the private key is used as both a private key-transport key to decrypt data-encryption keys and a private signature key to apply digital signatures. It may be necessary to retain the private key-transport key beyond the cryptoperiod of the corresponding public key-transport key in order to decrypt the data-encryption keys needed to access encrypted data. On the other hand, the private signature key **shall** be destroyed at the expiration of its cryptoperiod to prevent its compromise (see Section 5.3.6). In this example, the longevity requirements for the private key-transport key and the private digital-signature key contradict each other.

This principle does not preclude using a single key in cases where the same process can provide multiple services. This is the case, for example, when a digital signature provides assurance of the identity of the originating entity, non-repudiation, source authentication and integrity protection using a single digital signature, or when a single symmetric data-encryption key can be used to encrypt and authenticate data in a single cryptographic operation (e.g., using an authenticated-encryption operation, as opposed to separate encryption and authentication operations). Also refer to Section 3.7.

This Recommendation also permits the use of a private key-transport or key-agreement key to generate a digital signature for the following special case:

When requesting the (initial) certificate for a static key-establishment key, the associated private key may be used to sign the certificate request. Also refer to Section 8.1.5.1.1.2.

5.3   Cryptoperiods

A cryptoperiod is the time span during which a specific key is authorized for use by legitimate entities, or the keys for a given system will remain in effect. A suitably defined cryptoperiod:

1. Limits the amount of information protected by a given key that is available for cryptanalysis,

2. Limits the amount of exposure if a single key is compromised,

3. Limits the use of a particular algorithm to its estimated effective lifetime,

4. Limits the time available for attempts to penetrate physical, procedural, and logical access mechanisms that protect a key from unauthorized disclosure,

5 Limits the period within which information may be compromised by inadvertent disclosure of keying material to unauthorized entities, and

6. Limits the time available for computationally intensive cryptanalytic attacks (in applications where long-term key protection is not required).

Sometimes cryptoperiods are defined by an arbitrary time period or maximum amount of data protected by the key. However, trade-offs associated with the determination of cryptoperiods involve the risk and consequences of exposure, which should be carefully considered when selecting the cryptoperiod (see Section 5.6.4).

5.3.1   Risk Factors Affecting Cryptoperiods

Among the factors affecting the risk of exposure are:

1. The strength of the cryptographic mechanisms (e.g., the algorithm, key length, block size, and mode of operation),

2. The embodiment of the mechanisms (e.g., a [FIPS140] Level 4 implementation or a software implementation on a personal computer),

3. The operating environment (e.g., a secure limited access facility, open office environment, or publicly accessible terminal),

4. The volume of information flow or the number of transactions,

5. The security life of the data,

6. The security function (e.g., data encryption, digital signature, key production or derivation, key protection),

7. The re-keying method (e.g., keyboard entry, re-keying using a key loading device where humans have no direct access to key information, remote re-keying within a PKI),

8. The key update or key-derivation process,

9. The number of nodes in a network that share a common key,

10. The number of copies of a key and the distribution of those copies,

11. Personnel turnover (e.g., CA system personnel), and

12. The threat to the information (e.g., whom the information is protected from, and what are their perceived technical capabilities and financial resources to mount an attack).

In general, short cryptoperiods enhance security. For example, some cryptographic algorithms might be less vulnerable to cryptanalysis if the adversary has only a limited amount of information encrypted under a single key. On the other hand, where manual key-distribution methods are subject to human error and frailty, more frequent key changes might actually increase the risk of exposure. In these cases, especially when very strong cryptography is employed, it may be more prudent to have fewer, well-controlled manual key distributions, rather than more frequent, poorly controlled manual key distributions.

In general, where strong cryptography is employed, physical, procedural, and logical access-protection considerations often have more impact on cryptoperiod selection than do algorithm and key-size factors. In the case of **approved** algorithms, modes of operation, and key sizes, adversaries may be able to access keys through penetration or subversion of a system with less expenditure of time and resources than would be required to mount and execute a cryptographic attack.

5.3.2 Consequence Factors Affecting Cryptoperiods

The consequences of exposure are measured by the sensitivity of the information, the criticality of the processes protected by the cryptography, and the cost of recovery from the compromise of the information or processes. Sensitivity refers to the lifespan of the information being protected (e.g., 10 minutes, 10 days or 10 years) and the potential consequences of a loss of protection for that information (e.g., the disclosure of the information to unauthorized entities). In general, as the sensitivity of the information or the criticality of the processes protected by cryptography increase, the length of the associated cryptoperiods **should** decrease in order to limit the damage that might result from each compromise. This is subject to the caveat regarding the security and integrity of the re-keying, key update or key-derivation process (see Sections 8.2.3 and 8.2.4). Short cryptoperiods may be counter productive, particularly where denial of service is the paramount concern, and there is a significant potential for error in the re-keying, key update or key-derivation process.

5.3.3 Other Factors Affecting Cryptoperiods

5.3.3.1 Communications versus Storage

Keys that are used for confidentiality protection of communications exchanges may often have shorter cryptoperiods than keys used for the protection of stored data. Cryptoperiods are generally made longer for stored data because the overhead of re-encryption associated with changing keys may be burdensome.

5.3.3.2 Cost of Key Revocation and Replacement

In some cases, the costs associated with changing keys are painfully high. Examples include decryption and subsequent re-encryption of very large databases, decryption and re-encryption of distributed databases, and revocation and replacement of a very large number of keys (e.g., where there are very large numbers of geographically and organizationally distributed key holders). In such cases, the expense of the security measures necessary to support longer cryptoperiods may be justified (e.g., costly and inconvenient physical, procedural, and logical access security; and the use of cryptography strong enough to support longer cryptoperiods even where this may result in significant additional processing overhead). In other cases, the cryptoperiod may be shorter than would otherwise be necessary; for example, keys may be changed frequently in order to limit the period of time that the key management system maintains status information.

5.3.4 Cryptoperiods for Asymmetric Keys

For key pairs, each key of the pair has its own cryptoperiod. That is, each key is used by an "originator" to apply cryptographic protection (e.g., create a digital signature) or by a "recipient" to subsequently process the protected information (e.g., verify a digital signature), but not both. Where public keys are distributed in public-key certificates, the cryptoperiod for each key of the key pair is not necessarily the same as the validity period of the certificate. The cryptoperiod of a public key is extended beyond the expiration date of a certificate when a new public-key certificate with the same subject public key and a later expiration date is issued.

See Section 5.3.6 for guidance regarding specific key types. Examples of cryptoperiod issues associated with public-key cryptography include:

1. The cryptoperiod of a private key-transport key may be longer than the cryptoperiod of the associated public key (i.e., the public key-transport key). The public key is used for a fixed period of time to encrypt keying material. That period of time may be indicated by the *expiration date* on a public-key certificate. The private key will need to be retained as long as there is a need to recover (i.e., decrypt) the key(s) encrypted by the public key.

2. In contrast, the cryptoperiod of a private authentication key that is used to sign challenge information is basically the same as the cryptoperiod of the associated public key (i.e., the public authentication key). That is, when the private key will not be used to sign challenges, the public key is no longer needed.

3. If a private signature key is used to generate digital signatures as a proof-of-origin, the cryptoperiod of the private key may be significantly shorter than the cryptoperiod of the associated public signature-verification key. In this case, the private key is usually intended for use for a fixed period of time, after which time the key owner **shall** destroy[11] the private key. The public key may be available for a longer period of time for verifying signatures. However, other factors, such as the strength of the signing algorithm, the value of the signature, and the likelihood of forgery, **should** be considered.

5.3.5   Symmetric Key Usage Periods and Cryptoperiods

For symmetric keys, a single key is used for both applying the protection (e.g., encrypting or computing a MAC) and processing the protected information (e.g., decrypting the encrypted information or verifying a MAC). The period of time during which cryptographic protection may be applied to data is called the *originator-usage period*, and the period of time during which the protected information is processed is called the *recipient-usage period*. A symmetric key **shall not** be used to provide protection after the end of the originator-usage period. The recipient-usage period may extend beyond the originator-usage period. This permits all information that has been protected by the originator to be processed by the recipient before the processing key is deactivated. However, in many cases, the originator and recipient-usage periods are the same. The (total) "cryptoperiod" of a symmetric key is the period of time from the beginning of the originator-usage period to the end of the recipient-usage period, although the originator-usage period has historically been used as the cryptoperiod for the key.

Note that in some cases, predetermined cryptoperiods may not be adequate for the security life of the protected data. If the required security life exceeds the cryptoperiod, then the protection will need to be reapplied using a new key.

Examples of the use of the usage periods include:

---

[11] A simple deletion of the keying material might not completely obliterate the information. For example, erasing the information might require overwriting that information multiple times with other non-related information, such as random bits, or all zero or one bits. Keys stored in memory for a long time can become "burned in". This can be mitigated by splitting the key into components that are frequently updated (see [DiCrescenzo]).

a. When a symmetric key is used only for securing communications, the period of time from the originator's application of protection to the recipient's processing is negligible. In this case, the key is authorized for either purpose during the entire cryptoperiod, i.e., the originator-usage period and the recipient-usage period are the same.

b. When a symmetric key is used to protect stored information, the originator-usage period (when the originator applies cryptographic protection to stored information) may end much earlier than the recipient-usage period (when the stored information is processed). In this case, the cryptoperiod begins at the initial time authorized for the application of protection with the key, and ends with the latest time authorized for processing using that key. In general, the recipient-usage period for stored information will continue beyond the originator-usage period, so that the stored information may be authenticated or decrypted at a later time.

c. When a symmetric key is used to protect stored information, the recipient-usage period may start after the beginning of the originator-usage period as shown in Figure 1. For example, information may be encrypted before being stored on a compact disk. At some later time, the key may be distributed in order to decrypt and recover the information.

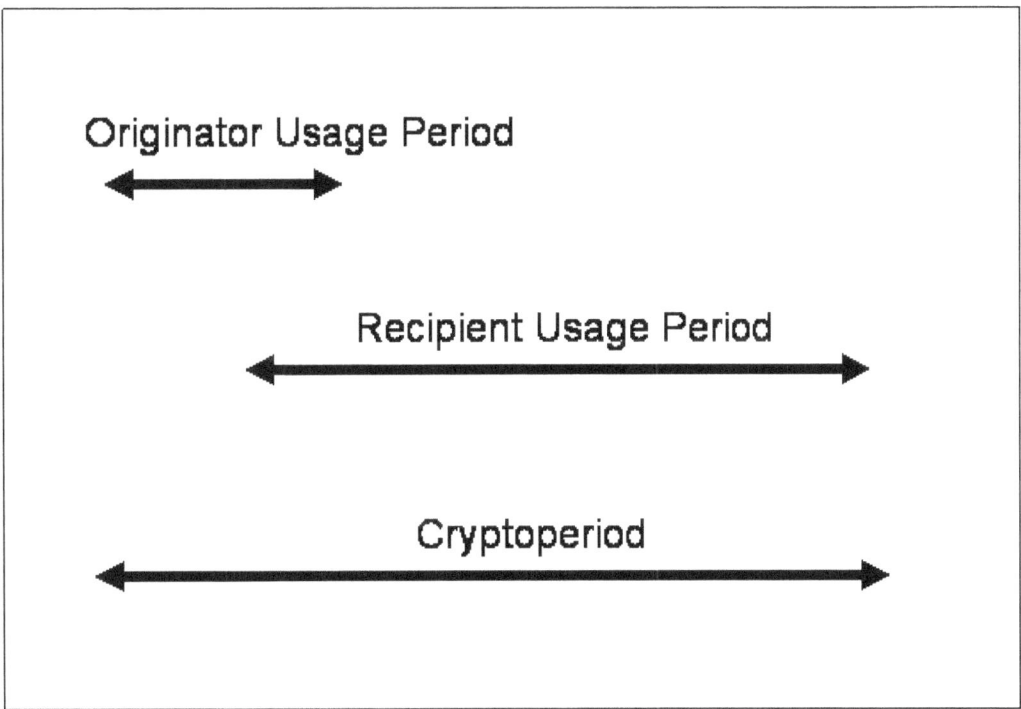

Figure 1: Symmetric key cryptoperiod (Example C)

5.3.6 Cryptoperiod Recommendations for Specific Key Types

The cryptoperiod required for a given key may be affected by key type as much as by the usage environment and data characteristics described above. Some general cryptoperiod recommendations for various key types are suggested below. Note that the cryptoperiods suggested are only rough order-of-magnitude guidelines; longer or shorter cryptoperiods may be warranted, depending on the application and environment in which the keys will be used.

However, when assigning a longer cryptoperiod than that suggested below, serious consideration should be given to the risks associated with doing so (see Section 5.3.1). Most of the suggested cryptoperiods are on the order of 1-2 years, based on 1) a desire for maximum operational efficiency and 2) assumptions regarding the minimum criteria for the usage environment (see [FIPS140], [SP800-14], [SP800-21], and [SP800-37]). The factors described in Sections 5.3.1 through 5.3.3 **should** be used to determine actual cryptoperiods for specific usage environments.

1. *Private signature key*:

    a. Type Considerations: In general, the cryptoperiod of a private signature key may be shorter than the cryptoperiod of the corresponding public signature-verification key.

    b. Cryptoperiod: Given the use of **approved** algorithms and key sizes, and an expectation that the security of the key-storage and use environment will increase as the sensitivity and/or criticality of the processes for which the key provides integrity protection increases, a maximum cryptoperiod of about 1-3 years is recommended. The key **shall** be destroyed at the end of its cryptoperiod.

2. *Public signature-verification key*:

    a. Type Considerations: In general, the cryptoperiod of a public signature-verification key may be longer than the cryptoperiod of the corresponding private signature key. The cryptoperiod is, in effect, the period during which any signature computed using the associated private signature key needs to be verified. A longer cryptoperiod for the public signature-verification key (than the private signature key) poses a relatively minimal security concern.

    b. Cryptoperiod: The cryptoperiod may be on the order of several years, though due to the long exposure of protection mechanisms to hostile attack, the reliability of the signature is reduced with the passage of time. That is, for any given algorithm and key size, vulnerability to cryptanalysis is expected to increase with time. Although choosing the strongest available algorithm and a large key size can minimize this vulnerability to cryptanalysis, the consequences of exposure to attacks on physical, procedural, and logical access-control mechanisms for the private key are not affected.

    Some systems use a cryptographic timestamping function to place an unforgeable timestamp on each signed message. These systems can have a public signature-verification key cryptoperiod that is about the same as the private signature key cryptoperiod. Even though the cryptoperiod has expired, the public signature-verification key may be used to validate signatures on messages whose timestamps are within the cryptoperiod of the verification key. In this case, one is relying on the cryptographic timestamp function to assure that the message was signed within its cryptoperiod.

3. *Symmetric authentication key*:

    a. Type Considerations: The cryptoperiod of a symmetric authentication key depends on the sensitivity of the type of information it protects and the protection afforded the key. For very sensitive information, the authentication key may need to be unique to the protected information. For less sensitive information, suitable cryptoperiods may extend beyond a single use. The originator-usage period of a symmetric authentication key applies to the use of that key in applying the original cryptographic protection for the information (e.g., computing the MAC to be associated with the authenticated information); new MACs **shall not** be computed on information using that key after the end of the originator-usage period. However, the key may need to be available to verify the MAC on the protected data beyond the originator-usage period (i.e., the recipient-usage period extends beyond the originator-usage period). Note that if a MAC key is compromised, it may be possible for an adversary to modify the data that was authenticated and then recalculate the MAC.

b. Cryptoperiod: Given the use of **approved** algorithms and key sizes and an expectation that the security of the key-storage and use environment will increase as the sensitivity and/or criticality of the processes for which the key provides integrity protection increases, a maximum originator-usage period of up to 2 years is recommended, and a maximum recipient-usage period of 3 years beyond the end of the originator-usage period is recommended.

4. *Private authentication key*:

   a. Type Considerations: A private authentication key may be used multiple times. Its associated public key could be certified, for example, by a Certification Authority. In most cases, the cryptoperiod of the authentication private key is the same as the cryptoperiod of the associated public key.

   b. Cryptoperiod: An appropriate cryptoperiod for a private authentication key would be 1-2 years, depending on its usage environment and the sensitivity/criticality of the authenticated information.

5. *Public authentication key*:

   a. Type Considerations: In most cases, the cryptoperiod of a public authentication key is the same as the cryptoperiod of the associated private authentication key. The cryptoperiod is, in effect, the period during which the identity of the originator of information protected by the associated authentication private key needs to be verified.

   b. Cryptoperiod: An appropriate cryptoperiod for the authentication public key would be 1-2 years, depending on its usage environment and the sensitivity/criticality of the authenticated information.

6. *Symmetric data-encryption key*:

   a. Type Considerations: A symmetric data-encryption key is used to protect stored data, messages or communications sessions. Based primarily on the consequences of compromise, a data-encryption key that is used to encrypt large volumes of information over a short period of time (e.g., for link encryption) **should** have a relatively short originator-usage period. An encryption key used to encrypt less information could have a longer originator-usage period. The originator-usage period of a symmetric data-encryption key applies to the use of that key in applying the original cryptographic protection for information (i.e., encrypting the information) (see Section 5.3.5).

   During the originator-usage period, information may be encrypted by the data-encryption key; the key **shall not** be used for performing an encryption operation on information beyond this period. However, the key may need to be available to decrypt the protected data beyond the originator-usage period (i.e., the recipient-usage period may need to extend beyond the originator-usage period).

b. Cryptoperiod: The originator-usage period recommended for the encryption of large volumes of information over a short period of time (e.g., for link encryption) is on the order of a day or a week. An encryption key used to encrypt smaller volumes of information might have an originator-usage period of up to one month. A maximum recipient-usage period of 3 years beyond the end of the originator-usage period is recommended.

In the case of symmetric data-encryption keys that are used to encrypt single messages or single communications sessions, the lifetime of the protected data could be months or years because the encrypted messages may be stored for later reading. Where information is maintained in encrypted form, the symmetric data-encryption keys need to be maintained until that information is re-encrypted under a new key or destroyed. Note that confidence in the confidentiality of the information is reduced with the passage of time.

7. *Symmetric key-wrapping key*:

a. Type Considerations: A symmetric key-wrapping key that is used to encrypt very large numbers of keys over a short period of time **should** have a relatively short originator-usage period. If a small number of keys are encrypted, the originator-usage period of the key-wrapping key could be longer. The originator-usage period of a symmetric key-wrapping key applies to the use of that key in providing the original protection for information (i.e., encrypting the key that is to remain secret); keys **shall not** be encrypted using the key-wrapping key after the end of the originator-usage period. However, the key may need to be available to decrypt the protected data beyond the originator-usage period (i.e., the recipient-usage period may need to extend beyond the originator-usage period).

Some symmetric key-wrapping keys are used for only a single message or communications session. In the case of these very short-term key-wrapping keys, an appropriate cryptoperiod (i.e., which includes both the originator and recipient-usage periods) is a single communication session. It is assumed that the key as encrypted by the key-wrapping key will not be retained in its encrypted form, so the originator-usage period of the key-wrapping key as used for encryption is the same as the recipient-usage period of that key when used for decryption. In other cases, key-wrapping keys may be retained so that the files or messages encrypted by the wrapped keys may be recovered later on. In this case the recipient-usage period may be significantly longer than the originator-usage period, and cryptoperiods lasting for years may be employed.

b. Cryptoperiod: The recommended originator-usage period for a symmetric key-wrapping key that is used to encrypt very large numbers of keys over a short period of time is on the order of a day or a week. If a relatively small number of keys are to be encrypted under the key-wrapping key, the originator-usage period of the key-wrapping key could be up to a month. In the case of keys used for only a single message or communications session, the cryptoperiod would be limited to a single communication session. Except for the latter, a maximum recipient-usage period of 3 years beyond the end of the originator-usage period is recommended.

8. *Symmetric and Asymmetric RNG keys*:

a. Type Considerations: Symmetric and asymmetric RNG keys are used in deterministic random number generation functions. The **approved** RNGs in [SP800-90A] control key changes (e.g., during reseeding).

b. Cryptoperiod: Assuming the use of **approved** RNGs, the maximum cryptoperiod of symmetric and asymmetric RNG keys is determined by the design of the RNG.

9. *Symmetric master key*:

a. Type Considerations: A symmetric master key may be used multiple times to derive other keys using a (one-way) key-derivation function (see Section 8.2.4). Therefore, the cryptoperiod consists of only an originator-usage period for this key type. A suitable cryptoperiod depends on the nature and use of the keys derived from the master key and on considerations provided earlier in Section 5.3. The cryptoperiod of a key derived from a master key could be relatively short, e.g., a single use, communication session, or transaction. Alternatively, the master key could be used over a longer period of time to derive (or re-derive) multiple keys for the same or different purposes. The cryptoperiod of the derived keys depends on their use (e.g., as symmetric data-encryption or authentication keys).

b. Cryptoperiod: An appropriate cryptoperiod for the symmetric master key might be 1 year, depending on its usage environment and the sensitivity/criticality of the information protected by the derived keys and the number of keys derived from the master key.

10. *Private key-transport key*:

a. Type Considerations: A private key-transport key may be used multiple times. Due to the potential need to decrypt keys some time after they have been encrypted for transport, the cryptoperiod of the private key-transport key may be longer than the cryptoperiod of the associated public key. The cryptoperiod of the private key is the length of time during which any keys encrypted by the associated public key-transport key need to be decrypted.

b. Cryptoperiod: Given 1) the use of **approved** algorithms and key sizes, 2) the volume of information that may be protected by keys encrypted under the associated public key-transport key, and 3) an expectation that the security of the key-storage and use environment will increase as the sensitivity and/or criticality of the processes for which the key provides protection increases; a maximum cryptoperiod of about 2 years is recommended. In certain applications (e.g., email), where received messages are stored and decrypted at a later time, the cryptoperiod of the private key-transport key may exceed the cryptoperiod of the public key-transport key.

11. *Public key-transport key*:

a. Type Considerations: The cryptoperiod for the public key-transport key is that period of time during which the public key may be used to actually apply the encryption operation to the keys that will be protected. Public key-transport keys can be publicly known. The driving factor in establishing the public key-transport key cryptoperiod is the cryptoperiod of the associated private key-transport key. As indicated in the private key-transport key discussion, due to the potential need to decrypt keys some time after they have been encrypted for transport, the cryptoperiod of the public key-transport key may be shorter than that of the associated private key.

b. Cryptoperiod: Based on cryptoperiod assumptions for associated private keys, a recommendation for the maximum cryptoperiod might be about 1 - 2 years.

12. *Symmetric key-agreement key*:

a. Type Considerations: A symmetric key-agreement key may be used multiple times. Generally, the originator-usage period and the recipient-usage period are the same. The cryptoperiod of these keys depends on 1) environmental security factors, 2) the nature (e.g., types and formats) and volume of keys that are established, and 3) the details of the key-agreement algorithms and protocols employed. Note that symmetric key-agreement keys may be used to establish symmetric keys (e.g., symmetric data encryption keys) or other keying material (e.g., IVs).

b. Cryptoperiod: Given an assumption that the cryptography that employs symmetric key-agreement keys 1) employs an **approved** algorithm and key scheme, 2) the cryptographic device meets [FIPS140] requirements, and 3) the risk levels are established in conformance to [FIPS199], an appropriate cryptoperiod for the key would be 1-2 years.

13. *Private static key-agreement key*:

a. Type Considerations: A private static key-agreement key may be used multiple times. As in the case of symmetric key-agreement keys, the cryptoperiod of these keys depends on 1) environmental security factors, 2) the nature (e.g., types and formats) and volume of keys that are established, and 3) the details of the key-agreement algorithms and protocols employed. Note that private static key-agreement keys may be used to establish symmetric keys (e.g., key-wrapping keys) or other secret keying material.

b. Cryptoperiod: Given an assumption that the cryptography that employs private static key-agreement keys 1) employs an **approved** algorithm and key scheme, 2) the cryptographic device meets [FIPS140] requirements, and 3) the risk levels are established in conformance to [FIPS199], an appropriate cryptoperiod for the key would be 1-2 years. In certain applications (e.g., email), where received messages are stored and decrypted at a later time, the cryptoperiod of the private static key-agreement key may exceed the cryptoperiod of the public static key-agreement key associated with the private key .

14. *Public static key-agreement key*:

a. Type Considerations: The cryptoperiod for a public static key-agreement key is usually the same as the cryptoperiod of the associated private static key-agreement key. See the discussion for the private static key-agreement key.

b. Cryptoperiod: The cryptoperiod of the public static key-agreement key may be 1-2 years.

15. *Private ephemeral key-agreement key*:

a. Type Considerations: Private ephemeral key-agreement keys are the private key elements of asymmetric key pairs that are used in a single transaction to establish one or more keys. Private ephemeral key-agreement keys may be used to establish symmetric keys (e.g., key-wrapping keys) or other secret keying material.

b. Cryptoperiod: Private ephemeral key-agreement keys are used for a single key-agreement transaction. However, a private ephemeral key may be used multiple times to establish the same symmetric key with multiple parties during the same transaction (broadcast). The cryptoperiod of a private ephemeral key-agreement key is the duration of a single key-agreement transaction.

16. *Public ephemeral key-agreement key*:

a. Type Considerations: Public ephemeral key-agreement keys are the public key elements of asymmetric key pairs that are used only once to establish one or more keys.

b. Cryptoperiod: Public ephemeral key-agreement keys are used for a single key-agreement transaction. The cryptoperiod of the public ephemeral key-agreement key ends immediately after it is used to generate the shared secret. Note that in some cases, the cryptoperiod of the public ephemeral key-agreement key may be different for the participants in the key-agreement transaction. For example, consider an encrypted email application in which the email sender generates an ephemeral key-agreement key pair, and then uses the key pair to generate an encryption key that is used to encrypt the contents of the email. For the sender, the cryptoperiod of the public key ends when the shared secret is generated and the *encryption* key is derived. However, for the encrypted email receiver, the cryptoperiod of the ephemeral public key does not end until the shared secret is generated and the *decryption* key is derived; if the email is not processed immediately upon receipt (e.g., it is decrypted a week later than the email was sent), then the cryptoperiod of the ephemeral public key does not end (from the perspective of the receiver) until the shared secret is generated that uses that public key.

17. *Symmetric authorization key*:

a. Type Considerations: A symmetric authorization key may be used for an extended period of time, depending on the resources that are protected and the role of the entity authorized for access. For this key type, the originator-usage period and the recipient-usage period are the same. Primary considerations in establishing the cryptoperiod for symmetric authorization keys include the robustness of the key, the adequacy of the cryptographic method, and the adequacy of key-protection mechanisms and procedures.

b. Cryptoperiod: Given the use of **approved** algorithms and key sizes, and an expectation that the security of the key-storage and use environment will increase as the sensitivity and criticality of the authorization processes increases, it is recommended that cryptoperiods be no more than two years.

18. *Private authorization key*:

a. Type Considerations: A private authorization key may be used for an extended period of time, depending on the resources that are protected and the role of the entity authorized for access. Primary considerations in establishing the cryptoperiod for private authorization keys include the robustness of the key, the adequacy of the cryptographic method, and the adequacy of key-protection mechanisms and procedures. The cryptoperiod of the private authorization key and its associated public key **shall** be the same.

b. Cryptoperiod: Given the use of **approved** algorithms and key sizes, and an expectation that the security of the key-storage and use environment will increase as the sensitivity and criticality of the authorization processes increases, it is recommended that cryptoperiods for private authorization keys be no more than two years.

19. *Public authorization key*:

a. Type Considerations: A public authorization key is the public element of an asymmetric key pair used to verify privileges for an entity that possesses the associated private key.

b. Cryptoperiod: The cryptoperiod of the public authorization key **shall** be the same as the authorization private key: no more than two years.

Table 1 below is a summary of the cryptoperiods that are suggested for each key type. Longer or shorter cryptoperiods may be warranted, depending on the application and environment in which the keys will be used. However, when assigning a longer cryptoperiod than that suggested below, serious consideration **should** be given to the risks associated with doing so (see Section 5.3.1).

Table 1: Suggested cryptoperiods for key types[12]

Key Type	Cryptoperiod	
	Originator Usage Period (OUP)	Recipient Usage Period
1. Private Signature Key	1-3 years	
2. Public Signature Key	Several years (depends on key size)	
3. Symmetric Authentication Key	$\leq$ 2 years	$\leq$ OUP + 3 years
4. Private Authentication Key	1-2 years	
5. Public Authentication Key	1-2 years	
6. Symmetric Data Encryption Keys	$\leq$ 2 years	$\leq$ OUP + 3 years
7. Symmetric Key Wrapping Key	$\leq$ 2 years	$\leq$ OUP + 3 years
8. Symmetric and asymmetric RNG Keys	Upon reseeding	
9. Symmetric Master Key	About 1 year	
10. Private Key Transport Key	$\leq$ 2 years[13]	
11. Public Key Transport Key	1-2 years	
12. Symmetric Key Agreement Key	1-2 years	
13. Private Static Key Agreement Key	1-2 years[14]	

---

[12] In some cases, risk factors affect the cryptoperiod selection (see Section 5.3.1).

[13] In certain email applications where received messages are stored and decrypted at a later time, the cryptoperiod of the private key-transport key may exceed the cryptoperiod of the public key-transport key.

[14] In certain email applications whereby received messages are stored and decrypted at a later time, the cryptoperiod of the private static key-agreement key may exceed the cryptoperiod of the public static key-agreement key.

Key Type	Cryptoperiod	
	Originator Usage Period (OUP)	**Recipient Usage Period**
14. Public Static Key Agreement Key	1-2 years	
15. Private Ephemeral Key Agreement Key	One key-agreement transaction	
16. Public Ephemeral Key Agreement Key	One key-agreement transaction	
17. Symmetric Authorization Key	$\leq$ 2 years	
18. Private Authorization Key	$\leq$ 2 years	
19. Public Authorization Key	$\leq$ 2 years	

5.3.7   Recommendations for Other Keying Material

Other keying material does not have well-established cryptoperiods, per se. The following recommendations are offered regarding the disposition of this other keying material:

1. Domain parameters remain in effect until changed.

2. An IV is associated with the information that it helps to protect, and is needed until the information and its protection are no longer needed.

3. Shared secrets generated during the execution of key-agreement schemes **shall** be destroyed as soon as they are no longer needed to derive keying material.

4. RNG seeds **shall** be destroyed immediately after use.

5. Other public information **should not** be retained longer than needed for cryptographic processing.

6. Other secret information **shall not** be retained longer than necessary.

7. Intermediate results **shall** be destroyed immediately after use.

5.4   Assurances

When cryptographic keys and domain parameters are stored or distributed they may pass through unprotected environments. In this case, specific assurances may be required before the key or domain parameters may be used to perform normal cryptographic operations.

5.4.1   Assurance of Integrity (Also Integrity Protection)

Assurance of integrity **shall** be obtained prior to using all keying material.

At a minimum, assurance of integrity **shall** be obtained by verifying that the keying material has the appropriate format and came from an authorized source. Additional assurance of integrity may be obtained by the proper use of error detection codes, message authentication codes, and digital signatures.

5.4.2   Assurance of Domain Parameter Validity

Domain parameters are used by some public-key algorithms during the generation of key pairs and digital signatures, and during the generation of shared secrets (during the execution of a key-agreement scheme) that are subsequently used to derive keying material. Assurance of the validity of the domain parameters is important to applications of public-key cryptography and **shall** be obtained prior to using them.

Invalid domain parameters could void all intended security for all entities using the domain parameters. Methods of obtaining assurance of domain-parameter validity for DSA, and finite-field discrete-log key-agreement algorithms are provided in [SP800-89] and [SP800-56A]. Methods for obtaining this assurance for ECDSA, and the elliptic-curve discrete-log key-establishment algorithms are provided in [SP800-56A].

5.4.3   Assurance of Public-Key Validity

Assurance of public-key validity **shall** be obtained on all public keys before using them.

Assurance of public-key validity gives the user confidence that the public key is arithmetically correct. This reduces the probability of using weak or corrupted keys. Invalid public keys could result in voiding the intended security, including the security of the operation (i.e., digital signature, key establishment, encryption), leaking some or all information from the owner's private key, and leaking some or all information about a private key that is combined with an invalid public key (as may be done when key agreement or public-key encryption is performed). One of several ways to obtain assurance of validity is to verify certain mathematical properties that the public key should have. Another way is to obtain the assurance from a trusted third party that the trusted party validated the properties.

Methods of obtaining assurance of public-key validity for DSA, and finite-field discrete-log key-agreement algorithms are provided in [SP800-89] and [SP800-56A]. Methods for obtaining this assurance for ECDSA, and the elliptic-curve discrete-log key-establishment algorithms are provided in [SP800-56A]. Methods for obtaining assurance of (partial) public-key validity for RSA are provided in [SP800-89], [SP800-56B] and [ANSX9.44].

5.4.4   Assurance of Private-Key Possession

Assurance of static private-key possession **shall** be obtained before the use of the corresponding static public key. Assurance of validity **shall** always be obtained prior to, or concurrently with, assurance of possession. Assurance of private-key possession **shall** be obtained by both the owner of the key pair and by other entities that receive the public key of that key pair and use it to interact with the owner.

The owner of the key pair obtains assurance of private key- possession by:

- Generating the key pair, or
- Performing a pair-wise consistency test (e.g., using the static private signature key to generate a digital signature, followed by a verification of the digital signature using the static public signature-verification key). Note that the key pair may have been generated by a trusted party and provided to the owner; the pair-wise consistency test will verify that the correct private key has been provided to the owner.

For parties other than the key pair owner, assurance of private-key possession gives confidence that the claimed owner of the public key actually possessed the corresponding private key at some time. There are several ways of obtaining assurance of private-key possession, in this case. The assurance may be obtained by participating in a protocol with the claimed owner of the key that uses the private key as it is intended to be used. For example, a private digital-signature key may be confirmed by using it to sign data (see Section 8.1.5.1.1.1, item 1), and a private key-establishment key may be confirmed by performing a key-confirmation protocol with the claimed owner of the key (see [SP800-56A] and [SP800-56B]). In the case of key-establishment keys, assurance of private-key possession may be obtained using the private key to digitally sign a certificate request (see Section 8.1.5.1.1.2). Sometimes when a CA public key is distributed, the CA will sign its own public key to provide assurance of private-key possession.

For specific details regarding assurance of the possession of private key-establishment keys, see [SP800-56A] and [SP800-56B]; for specific details regarding assurance of the possession of private digital-signature keys, see [SP800-89].

5.5   Compromise of Keys and other Keying Material

Information protected by cryptographic mechanisms is secure only if the algorithms remain strong, and the keys have not been compromised. Key compromise occurs when the protective mechanisms for the key fail (e.g., the confidentiality, integrity or association of the key to its owner fail - see Section 6), and the key can no longer be trusted to provide the required security. When a key is compromised, all use of the key to apply cryptographic protection to information (e.g., compute a digital signature or encrypt information) **shall** cease, and the compromised key **shall** be revoked (see Section 8.3.5). However, the continued use of the key under controlled circumstances to remove or verify the protections (e.g., decrypt or verify a digital signature) may be warranted, depending on the risks of continued use and an organization's Key Management Policy (see Part 2). The continued use of a compromised key **shall** be limited to processing protected information. In this case, the entity that uses the information **shall** be made fully aware of the dangers involved. Limiting the cryptoperiod of the key limits the amount of material that would be compromised (exposed) if the key were compromised. Using different keys for different purposes (e.g., different applications as well as different cryptographic mechanisms), as well as limiting the amount of information protected by a single key, also achieves this purpose.

The compromise of a key has the following implications:

1. The unauthorized disclosure of a key means that another entity (an unauthorized entity) may know the key and be able to use that key to perform computations requiring the use of the key.

    In general, the unauthorized disclosure of a key used to provide confidentiality protection[15] (i.e., via encryption) means that all information encrypted by that key could be known by unauthorized entities. For example, if a symmetric data-encryption key is compromised, the unauthorized entity might use the key to decrypt past or future encrypted information, i.e., the information is no longer confidential between the authorized entities.

---

[15] As opposed to the confidentiality of a key that could, for example, be used as a signing private key.

In the case of the unauthorized disclosure of a key used to provide integrity protection (e.g., via digital signatures), the integrity protection on the data may be lost. For example, if a private signature key is compromised, the unauthorized entity might sign messages as if they were originated by the key's real owner (either new messages or messages that are altered from their original contents), i.e., non-repudiation and the authenticity of the information are in question.

The unauthorized disclosure of a private signature key means that the integrity and non-repudiation qualities of all data signed by that key are suspect. An unauthorized party in possession of the private key could sign false information and make it appear to be valid. In cases where it can be shown that the signed data was protected by other mechanisms (e.g., physical security) from a time before the compromise, the signature may still have some value. For example, if a signed message was received on day 1, and it was later determined that the private signing key was compromised on day 15, the receiver may still have confidence that the message is valid because it was maintained in the receiver's possession. Note that cryptographic timestamping may also provide protection for messages signed before the private signature key was compromised. However, the security provided by these other mechanisms is now critical to the security of the signature. In addition, the non-repudiation of the signed message may be questioned, since the private signature key may have been disclosed to the message receiver, who then altered the message in some way.

The disclosure of a CA's private signature key means that an adversary can create fraudulent certificates and Certificate Revocation Lists (CRLs).

2. A compromise of the integrity of a key means that the key is incorrect - either that the key has been modified (either deliberately or accidentally), or that another key has been substituted; this includes a deletion (non-availability) of the key. The substitution or modification of a key used to provide integrity[16] calls into question the integrity of all information protected by the key. This information could have been provided by, or changed by, an unauthorized entity that knows the key. The substitution of a public or secret key that will be used (at a later time) to encrypt data could allow an unauthorized entity (who knows the decryption key) to decrypt data that was encrypted using the encryption key.

3. A compromise of a key's usage or application association means that the key could be used for the wrong purpose (e.g., for key establishment instead of digital signatures) or for the wrong application, and could result in the compromise of information protected by the key.

4. A compromise of a key's association with the owner or other entity means that the identity of the other entity cannot be assured (i.e., one does not know who the other entity really is) or that information cannot be processed correctly (e.g., decrypted with the correct key).

5. A compromise of a key's association with other information means that there is no association at all, or the association is with the wrong "information". This could cause the

---

[16] As opposed to the integrity of a key that could, for example, be used for encryption.

cryptographic services to fail, information to be lost, or the security of the information to be compromised.

Certain protective measures may be taken in order to minimize the likelihood or consequences of a key compromise. The following procedures are usually involved:

a. Limiting the amount of time a symmetric or private key is in plaintext form.

b. Preventing humans from viewing plaintext symmetric and private keys.

c. Restricting plaintext symmetric and private keys to physically protected containers. This includes key generators, key-transport devices, key loaders, cryptographic modules, and key-storage devices.

d. Using integrity checks to ensure that the integrity of a key or its association with other data has not been compromised. For example, keys may be wrapped (i.e., encrypted) in such a manner that unauthorized modifications to the wrapping or to the associations will be detected.

e. Employing key confirmation (see Section 4.2.5.5) to help ensure that the proper key was, in fact, established.

f. Establishing an accountability system that keeps track of each access to symmetric and private keys in plaintext form.

g. Providing a cryptographic integrity check on the key (e.g., using a MAC or a digital signature).

h. The use of trusted timestamps for signed data.

i. Destroying keys as soon as they are no longer needed.

j. Creating a compromise-recovery plan, especially in the case of a CA compromise.

The worst form of key compromise is one that is not detected. Nevertheless, even in this case, certain protective measures can be taken. Key management systems (KMSs) **should** be designed to mitigate the negative effects of a key compromise. A KMS **should** be designed so that the compromise of a single key compromises as little data as possible. For example, a single cryptographic key could be used to protect the data of only a single user or a limited number of users, rather than a large number of users. Often, systems have alternative methods to authenticate communicating entities that do not rely solely on the possession of keys. The object is to avoid building a system with catastrophic weaknesses.

A compromise-recovery plan is essential for restoring cryptographic security services in the event of a key compromise. A compromise-recovery plan **shall** be documented and easily accessible. The plan may be included in the Key Management Practices Statement (see Part 2). If not, the Key Management Practices Statement **should** reference the compromise-recovery plan.

Although compromise recovery is primarily a local action, the repercussions of a key compromise are shared by the entire community that uses the system or equipment. Therefore, compromise-recovery procedures **should** include the community at large. For example, recovery from the compromise of a root CA's private signature key requires that all users of the infrastructure obtain and install a new trust anchor. Typically, this involves physical procedures that are expensive to implement. To avoid these expensive procedures, elaborate precautions to avoid compromise may be justified.

The compromise-recovery plan **should** contain:

1. The identification of the personnel to notify,

2. The identification of the personnel to perform the recovery actions,

3. The re-key method,

4. An inventory of all cryptographic keys (e.g., the location of all certificates in a system),

5. The education of all appropriate personnel on the recovery procedures,

6. An identification of all personnel needed to support the recovery procedures,

7. Policies that key-revocation checking be enforced (to minimize the effect of a compromise),

8. The monitoring of the re-keying operations (to ensure that all required operations are performed for all affected keys), and

9. Any other recovery procedures.

Other compromise-recovery procedures may include:

5. Physical inspection of the equipment,

6. Identification of all information that may be compromised as a result of the incident,

7. Identification of all signatures that may be invalid, due to the compromise of a signing key, and

8. Distribution of new keying material, if required.

5.6   Guidance for Cryptographic Algorithm and Key-Size Selection

Cryptographic algorithms that provide the security services identified in Section 3 are specified in Federal Information Processing Standards (FIPS) and NIST Recommendations. Several of these algorithms are defined for a number of key sizes. This section provides guidance for the selection of appropriate algorithms and key sizes.

This section emphasizes the importance of acquiring cryptographic systems with appropriate algorithm and key sizes to provide adequate protection for 1) the expected lifetime of the system and 2) any data protected by that system during the expected lifetime of the data.

5.6.1   Comparable Algorithm Strengths

Cryptographic algorithms provide different "strengths" of security, depending on the algorithm and the key size used. In this discussion, two algorithms are considered to be of comparable strength for the given key sizes (X and Y) if the amount of work needed to "break the algorithms" or determine the keys (with the given key sizes) is approximately the same using a given resource. The security strength of an algorithm for a given key size is traditionally described in terms of the amount of work it takes to try all keys for a symmetric algorithm with a key size of "X" that has no short cut attacks (i.e., the most efficient attack is to try all possible keys). In this case, the best attack is said to be the exhaustion attack. An algorithm that has a Y-bit key, but whose strength is comparable to an X-bit key of such a symmetric algorithm is said have a "security strength of X bits" or to provide "X bits of security". Given a few plaintext blocks and corresponding ciphertext, an algorithm that provides X bits of security would, on average, take $2^{X-1}T$ units of time to attack, where T is the amount of time that is required to perform one encryption of a plaintext value and compare the result against the corresponding ciphertext value.

Determining the security strength of an algorithm can be nontrivial. For example, consider TDEA, which uses three 56-bit keys ($K1$, $K2$ and $K3$). If each of these keys is independently generated, then this is called the three-key option or three-key TDEA (3TDEA). However, if $K1$ and $K2$ are independently generated, and $K3$ is set equal to K1, then this is called the two-key option or two-key TDEA (2TDEA). One might expect that 3TDEA would provide $56 \times 3 = 168$ bits of strength. However, there is an attack on 3TDEA that reduces the strength to the work that would be involved in exhausting a 112-bit key. For 2TDEA, if exhaustion were the best attack, then the strength of 2TDEA would be $56 \times 2 = 112$ bits. This appears to be the case if the attacker has only a few matched plain and cipher pairs. However, the security strength of 2TDEA decreases as the number of matched plaintext/ciphertext pairs increases. If the attacker can obtain approximately 2^{40} such pairs, then 2TDEA has a security strength of about 80 bits.

The recommended, comparable key-size classes discussed in this section are based on assessments made as of the publication of this Recommendation using currently known methods. Advances in factoring algorithms, advances in general discrete-logarithm attacks, elliptic-curve discrete-logarithm attacks and quantum computing may affect these equivalencies in the future. New or improved attacks or technologies may be developed that leave some of the current algorithms completely insecure. If quantum attacks become practical, the asymmetric techniques may no longer be secure. Periodic reviews will be performed to determine whether the stated equivalencies need to be revised (e.g., the key sizes need to be increased) or the algorithms are no longer secure.

The use of strong cryptographic algorithms may mitigate security issues other than just brute-force cryptographic attacks. The algorithms may unintentionally be implemented in a manner that leaks small amounts of information about the key. In this case the larger key may reduce the likelihood that this leaked information will eventually compromise the key.

When selecting a block-cipher cryptographic algorithm (e.g., AES or TDEA), the block size may also be a factor that should be considered, since the amount of security provided by several of the modes defined in [SP800-38] is dependent on the block size[17]. More information on this issue is provided in [SP800-38].

Table 2 provides comparable security strengths for the **approved** algorithms; note that some of the larger key sizes are **not approved**.

1. Column 1 indicates the number of bits of security provided by the algorithms and key sizes in a particular row. Note that the number of bits of security is not necessarily the same as the key sizes for the algorithms in the other columns, due to attacks on those algorithms that provide computational advantages.

---

[17] Suppose that the block size is b bits. The collision resistance of a MAC is limited by the size of the Mactag, and collisions become probable after $2^{b/2}$ messages, if the full b bits are used as a Mactag. When using the Output Feedback mode of encryption, the maximum cycle length of the cipher can be at most 2^b blocks; the average cipher length is less than 2^b blocks. When using the Cipher Block Chaining mode, plaintext information is likely to begin to leak after $2^{b/2}$ blocks have been encrypted with the same key.

2. Column 2 identifies the symmetric-key algorithms that provide the indicated level of security (at a minimum), where 2TDEA and 3TDEA are specified in [SP800-67], and AES is specified in [FIPS197]. 2TDEA is TDEA with two different keys; 3TDEA is TDEA with three different keys.

3. Column 3 indicates the minimum size of the parameters associated with the standards that use finite-field cryptography (FFC). Examples of such algorithms include DSA as defined in [FIPS186] for digital signatures, and Diffie-Hellman (DH) and MQV key agreement as defined in [SP800-56A], where L is the size of the public key, and N is the size of the private key. The largest key size **approved** in [FIPS186] is ($L = 3072$, $N = 256$), and the largest key size **approved** in [SP800-56A] is ($L = 2048$, $N = 256$).

4. Column 4 indicates the value for k (the size of the modulus n) for algorithms based on integer-factorization cryptography (IFC). The predominant algorithm of this type is the RSA algorithm. RSA is specified in [ANSX9.31], [PKCS#1], [ANSX9.44] and [SP800-56B]. These specifications are referenced in [FIPS186] for digital signatures. The value of k is commonly considered to be the key size. The largest key size **approved** in [FIPS186] is $k = 3072$, and the largest key size **approved** in [SP800-56B] is $k = 2048$.

5. Column 5 indicates the range of f (the size of n, where n is the order of the base point G) for algorithms based on elliptic-curve cryptography (ECC) that are specified for digital signatures in [ANSX9.62] and adopted in [FIPS186], and for key establishment as specified in [SP800-56A]. The value of f is commonly considered to be the key size.

Table 2: Comparable strengths

Bits of security	Symmetric key algorithms	FFC (e.g., DSA, D-H)	IFC (e.g., RSA)	ECC (e.g., ECDSA)
80	2TDEA[18]	$L = 1024$ $N = 160$	$k = 1024$	$f = 160\text{-}223$
112	3TDEA	$L = 2048$ $N = 224$	$k = 2048$	$f = 224\text{-}255$
128	AES-128	$L = 3072$ $N = 256$	$k = 3072$	$f = 256\text{-}383$
192	AES-192	$L = 7680$ $N = 384$	$k = 7680$	$f = 384\text{-}511$
256	AES-256	$L = 15360$ $N = 512$	$k = 15360$	$f = 512+$

---

[18] The assessment of at least 80-bits of security for 2TDEA is based on the assumption that an attacker has no more than 2^{40} matched plaintext and ciphertext blocks (see [ANSX9.52], Annex B). Also see the example in the second paragraph of Section 5.6.1.

Appropriate hash functions that may be employed will be determined by the algorithm, scheme or application in which the hash function is used and by the minimum security-strength to be provided. Table 3 lists the hash functions that **shall** be used for providing the indicated security strength for the generation of digital signatures and HMAC values, for deriving keys using key-derivation functions (i.e., KDFs) and for random number generation. For some applications, a cryptographic key is associated with the application and needs to be considered when determining the security strength actually afforded by the application. For example, for the generation of digital signatures, the minimum key length for the keys for a given security strength is provided in the FFC, IFC and ECC columns of Table 2; while for HMAC, the key lengths are discussed in [SP800-107].

Table 3: Hash function that can be used to provide the targeted security strengths

Security Strength	Digital Signatures and hash-only applications	HMAC	Key Derivation Functions[19]	Random Number Generation[20]
80	SHA-1[21], SHA-224, SHA-512/224, SHA-256, SHA-512/256, SHA-384, SHA-512	SHA-1, SHA-512/224, SHA-224, SHA-256, SHA-512/256, SHA-384, SHA-512	SHA-1, SHA-224, SHA-512/224, SHA-256, SHA-512/256, SHA-384, SHA-512	SHA-1, SHA-224, SHA-512/224, SHA-256, SHA-512-/256, SHA-384, SHA-512
112	SHA-224, SHA-512/224, SHA-256, SHA-512/256, SHA-384, SHA-512	SHA-1, SHA-224, SHA-512/224, SHA-256, SHA-512/256, SHA-384, SHA-512	SHA-1, SHA-224, SHA-512/224, SHA-256, SHA-512/256, SHA-384, SHA-512	SHA-1, SHA-224, SHA-512/224, SHA-256, SHA-512/256, SHA-384, SHA-512
128	SHA-256, SHA-512/256, SHA-384, SHA-512	SHA-1, SHA-224, SHA-512/224, SHA-256,	SHA-1, SHA-224, SHA-512/224, SHA-256,	SHA-1, SHA-224, SHA-512/224, SHA-256,

---

[19] The security strength for key-derivation assumes that the shared secret contains sufficient entropy to support the desired security strength.

[20] The security strength assumes that the random number generator has been provided with adequate entropy to support the desired security strength.

[21] SHA-1 has been demonstrated to provide less than 80 bits of security for digital signatures; at the publication of this Recommendation, the security strength against collisions remains the subject of speculation. The use of SHA-1 is not recommended for the generation of digital signatures in new systems; new systems should use one of the larger hash functions. For the present time, SHA-1 is included here for digital signatures to reflect its widespread use in existing systems, for which the reduced security strength may not be of great concern when only 80 bits of security are required.

Security Strength	Digital Signatures and hash-only applications	HMAC	Key Derivation Functions[19]	Random Number Generation[20]
		SHA-512/256, SHA-384, SHA-512	SHA-512/256, SHA-384, SHA-512	SHA-512/256, SHA-384, SHA-512
192	SHA-384, SHA-512	SHA-224, SHA-512/224, SHA-256, SHA-512/256, SHA-384, SHA-512	SHA-224, SHA-512/224, SHA-256, SHA-512/256, SHA-384, SHA-512	SHA-224, SHA-512/224, SHA-256, SHA-512/256, SHA-384, SHA-512
256	SHA-512	SHA-256, SHA-512/256, SHA-384, SHA-512	SHA-256, SHA-512/256, SHA-384, SHA-512	SHA-256, SHA-512/256, SHA-384, SHA-512

5.6.2 Defining Appropriate Algorithm Suites

Many applications require the use of several different cryptographic algorithms. When several algorithms can be used to perform the same service, some algorithms are inherently more efficient because of their design (e.g., AES has been designed to be more efficient than Triple DEA).

In many cases, a variety of key sizes may be available for an algorithm. For some of the algorithms (e.g., public-key algorithms, such as RSA), the use of larger key sizes than are required may impact operations, e.g., larger keys may take longer to generate or longer to process the data. However, the use of key sizes that are too small may not provide adequate security.

Table 4 provides general recommendations that may be used to select an appropriate suite of algorithms and key sizes for Federal Government unclassified applications to protect sensitive data. A schedule for increasing the security strengths for applying cryptographic protection to data (e.g., encrypting or digitally signing) is specified in the table. Transition details for algorithms, key sizes and applications are provided in [SP800-131A]. The table is organized as follows:

1. Column 1 is divided into two sub-columns. The first sub-column indicates the security strength to be provided; the second sub-column indicates whether cryptographic protection is being applied to data (e.g., encrypted), or whether cryptographically protected data is being processed (e.g., decrypted).

2. Columns 2-5 indicate time frames during which the security strength is either acceptable, deprecated, OK for legacy use or deprecated. "Acceptable" indicates that the algorithm or key length is not known to be insecure. "Deprecated" means that the use of an algorithm

or key length that provides the indicated security strength may be used if risk is accepted; note that the use of deprecated algorithms or key lengths may have restrictions. "Legacy-use" means that an algorithm or key length may be used because of its use in legacy applications (i.e., the algorithm or key length can be used to process cryptographically protected data). "Disallowed" means that an algorithm or key length **shall not** be used for applying cryptographic protection. See [SP800-131A] for specific details and for any exceptions to the general guidance provided in Table 4 through 2015.

Table 4: Security-strength time frames

Security Strength		2011 through 2013	2014 through 2030	2031 and Beyond
80	Applying	Deprecated	Disallowed	
	Processing	Legacy use		
112	Applying	Acceptable	Acceptable	Disallowed
	Processing			Legacy use
128	Applying/Processing	Acceptable	Acceptable	Acceptable
192		Acceptable	Acceptable	Acceptable
256		Acceptable	Acceptable	Acceptable

If the security life of information extends beyond one time period specified in the table into the next time period (the later time period), the algorithms and key sizes specified for the later time period **shall** be used for applying cryptographic protection (e.g., encryption). The following examples are provided to clarify the use of the table:

1. If information is cryptographically protected (e.g., digitally signed) in 2012, and the maximum-expected security life of that data is only one year, any of the **approved** digital-signature algorithms or key sizes that provide at least 80 bits of security strength may be used. However, if only 80 bits of protection is used, there is some risk that needs to be accepted. Note that a digital signature that provides 80 bits of security could be processed (i.e., verified) after 2013 as indicated by the "legacy use" indication in the table.

2. If the information is to be digitally signed in 2012, and the expected security life of the data is six years, then an algorithm or key size that provides at least 112 bits of security strength is required.

5.6.3 Using Algorithm Suites

Algorithm suites that combine algorithms with a mixture of security strengths is generally discouraged. However, algorithms of different strengths and key sizes may be used together for performance, availability or interoperability reasons, provided that sufficient protection is provided. In general, the weakest algorithm and key size used to provide cryptographic protection determines the strength of the protection. Exceptions to this principle require extensive analysis. Determination of the strength of protection provided for information includes an analysis not only of the algorithm(s) and key size(s) used to apply the cryptographic protection(s) to the information, but also any algorithms and key sizes associated with establishing the key(s) used for information protection, including those used by communication protocols.

The following is a list of several algorithm combinations and discussions on the security implications of the combination:

1. When a key-establishment scheme is used to establish keying material for use with one or more algorithms (e.g., TDEA, AES, or HMAC), the strength of the selected combination is comparable to the weakest algorithm and key size used. For example, if a 224-bit ECC key is used to establish a 128-bit AES key (as defined in [SP800-56A]), only 112 bits of security are provided for any information protected by that AES key, since the 224-bit ECC provides only 112 bits of security. If 128 bits of security are required for the information protected by AES, then either an ECC key size of at least 256 bits, or another key-establishment algorithm with an appropriate key size needs to be selected to provide the required protection.

2. When a hash function and digital signature algorithm are used in combination to compute a digital signature, the strength of the signature is determined by the weaker of the two algorithms. For example, SHA-256 used with RSA using a 2048-bit key provides 112 bits of security, because a 2048-bit RSA key provides only 112 bits of security. If 128 bits of security is required, a 3072-bit RSA key would be appropriate.

3. When a random bit generator is used to generate a key for a cryptographic algorithm that is intended to provide X bits of security, an **approved** random bit generator **shall** be used that provides at least X bits of security.

If it is determined that a specific level of security is required for the protection of data, then an algorithm and key size suite needs to be selected that would provide that level of security as a minimum. For example, if 128 bits of security are required for data that is to be communicated and provided with confidentiality, integrity, authentication and non-repudiation protection, the following selection of algorithms and key sizes may be appropriate:

a. Confidentiality: Encrypt the information using AES-128. Other AES key sizes would also be appropriate, but performance may be slower.

b. Integrity, authentication and non-repudiation: Suppose that only one cryptographic operation is preferred. Use digital signatures. SHA-256 could be selected for the hash function. Select an algorithm for digital signatures from what is available to an application (e.g., ECDSA with at least a 256-bit key). If more than one algorithm and key size is available, the selection may be based on algorithm performance, memory requirements, etc., as long as the minimum requirements are met.

c. Key establishment: Select a key-establishment scheme that is based on the application and environment (see [SP800-56A] or [SP800-56B]), the availability of an algorithm in an implementation, and its performance. Select a key size from Table 2 for the algorithm that provides at least 128 bits of security. For example, if an ECC key-agreement scheme is available, use the ECC scheme with a 256-bit key. However, the key used for key agreement **shall** be different from the ECDSA key used for digital signatures.

Agencies that procure systems **should** consider the potential operational lifetime of the system. The agencies **shall** either select algorithms that are expected to be secure during the entire system lifetime, or **should** ensure that the algorithms and key sizes can be readily updated.

5.6.4 Transitioning to New Algorithms and Key Sizes

The estimated time period during which data protected by a specific cryptographic algorithm (and key size) remains secure is called the *algorithm security lifetime*. During this time, the algorithm may be used to both apply cryptographic protection (e.g., encrypt data) and to process the protected information (e.g., decrypt data); the algorithm is expected to provide adequate protection for the protected data during this period.

Typically, an organization selects the cryptographic services that are needed for a particular application. Then, based on the algorithm security lifetime and the security life of the data to be protected, an algorithm and key-size suite is selected that is sufficient to meet the requirements. The organization then establishes a key management system (if required), including validated cryptographic products that provide the services required by the application. As an algorithm and/or key-size suite nears its expiration date, transitioning to a new algorithm and key-size suite **should** be planned.

When the algorithm or key size is determined to no longer provide the desired protection for information (e.g., the algorithm may have been "broken"), any information "protected" by the algorithm or key size is considered to be "suspect" (e.g., the data may no longer be confidential, or the integrity cannot be assured). If the protected data is retained, it **should** be re-protected using an **approved** algorithm and key size that will protect the information for the remainder of its security life. However, it **should** be assumed that encrypted information could have been collected and retained by unauthorized entities (adversaries). The unauthorized entity may attempt to decrypt the information at some later time. In addition, the recovered plaintext could be used to attempt a matched plaintext-ciphertext attack on the new algorithm.

When using Table 2 and Table 4 to select the appropriate algorithm and key size, it is very important to take the expected security life of the data into consideration. As stated earlier, an algorithm (and key size) may be used both to apply cryptographic protection to data and process the protected data. When the security life of the data is taken into account, cryptographic protection **should not** be applied to data using a given algorithm (and key size) if the security life of the data extends beyond the end of the algorithm security lifetime (i.e., into the timeframe when the algorithm or key size is deprecated or disallowed; see Table 4). The period of time that an algorithm (and key size) may be used to apply cryptographic protection is called the *algorithm originator-usage period*. The algorithm security life = (the algorithm usage period + the security life of the data).

For example, suppose that 3TDEA was first used in January of 2010 for confidentiality protection in an application, and the security life of the data may be up to four years. Table 2 indicates that 3TDEA has a security strength of 112 bits. Table 4 indicates that an algorithm with a security strength of 112 bits has an algorithm security lifetime that extends through 2030 for applying cryptographic protection (i.e., encryption, in this case), but not beyond. However, since the data may have up to a four-year security life, the algorithm originator-usage period would have to end in 2026 rather than 2030 (i.e., the algorithm could not be used to encrypt data beyond 2026). See Figure 2. After 2026, the algorithm could be used to decrypt data for another four years, with the expectation that the confidentiality of the data continues to be protected at a security strength of 112 bits. If the security life of the data was estimated correctly, the data would no longer need this confidentiality protection. However, if the security life of the data is longer than originally expected, then the protection provided after 2030 may be less than required, and there is some risk that the confidentiality of the data may be compromised (after 2030); accepting the risk associated with the possible compromise is indicated by the "legacy use" indication in Table 4.

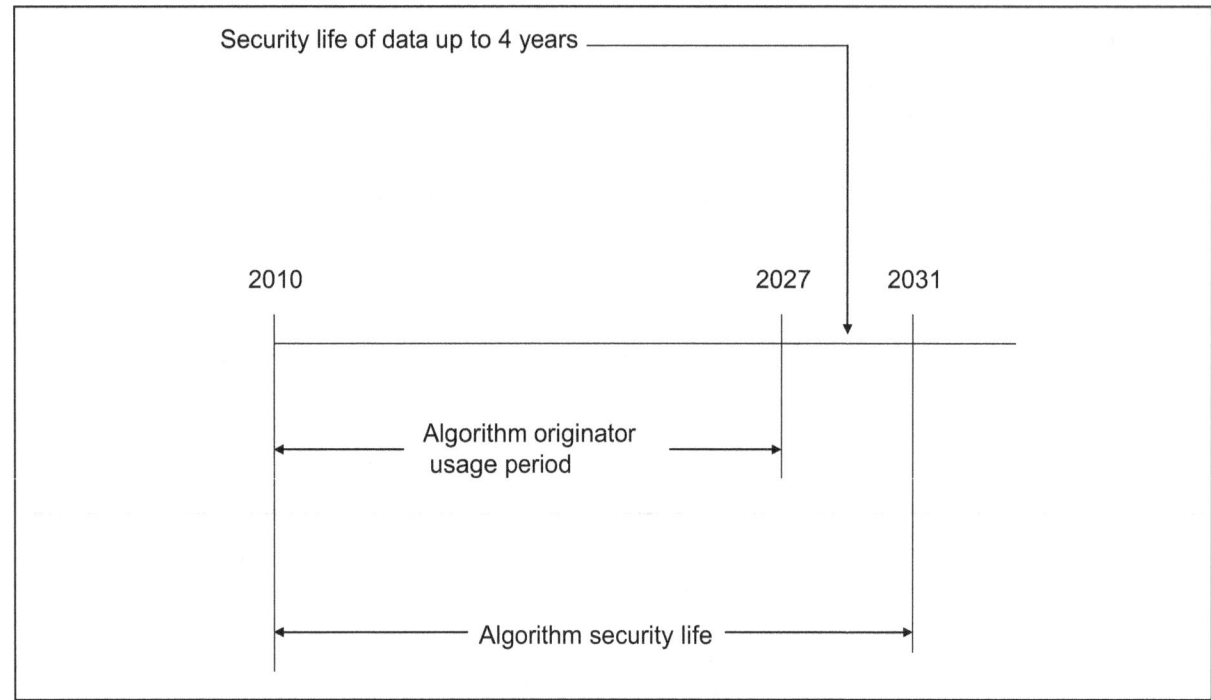

Figure 2: Algorithm Originator Usage Period Example

When initiating cryptographic protections for information, the strongest algorithm and key size that is appropriate for providing the protection **should** be used in order to minimize costly transitions. However, it should be noted that selecting some algorithms or key sizes that are unnecessarily large might have adverse performance effects (e.g., the algorithm may be unacceptably slow).

The process of transitioning to a new algorithm or a new key size may be as simple as selecting a more secure option in the security suites offered by the current system, or it can be as complex as building a whole new system. However, given that it is necessary to develop a new algorithm suite for a system, the following issues should be considered.

1. **Sensitivity of information and system lifetime:** The sensitivity of the information that will need to be protected by the system for the lifetime of the new algorithm(s) should be evaluated in order to determine the minimum security-requirement for the system. Care should be taken not to underestimate the lifetime of the system or the sensitivity of information that it may need to protect. Many decisions that were initially considered as temporary or interim decisions about data sensitivity have since been proven to be inadequate (e.g., the sensitivity of the information lasted well beyond its initially expected lifetime).

2. **Algorithm selection:** The new algorithms should be carefully selected to insure that they meet or exceed the minimum security-requirement of the system. In general, it is relatively easy to select cryptographic algorithms and key sizes that offer high security. However, it is wise for the amateur to consult a cryptographic expert when making such decisions. Systems **should** offer algorithm-suite options that provide for future growth.

3. **System design:** The new system **should** be designed to meet the minimum performance and security requirements. This is often a difficult task, since performance and security goals may conflict. All aspects of security (e.g., physical security, computer security, operational security, and personnel security) are involved. If a current system is to be modified to incorporate the new algorithms, the consequences need to be analyzed. For example, the existing system may require significant modifications to accommodate the footprints (e.g., key sizes, block sizes, etc.) of the new algorithms. In addition, the security measures (other than the cryptographic algorithms) retained from the current system **should** be reviewed to assure that they will continue to be effective in the new system.

4. **Pre-implementation evaluation:** Strong cryptography may be poorly implemented. Therefore, a changeover to new cryptographic techniques **should not** be made without an evaluation as to how effective and secure they are in the system.

5. **Testing:** Any complex system **should** be tested before it is employed.

6. **Training:** If the new system requires that new or different tasks (e.g., key management procedures) be performed, then the individuals who will perform those tasks **should** be properly trained. Features that are thought to be improvements may be viewed as annoyances by an untrained user.

7. **System implementation and transition:** Care **should** be taken to implement the system as closely as possible to the design. Exceptions **should** be noted.

8. **Transition:** A transition plan **should** be developed and followed so that the changeover from the old to the new system runs as smoothly as possible.

9. **Post-implementation evaluation**: The system **should** be evaluated to verify that the system as implemented meets the minimum security-requirements.

5.6.5   Security Strength Reduction

At some time, the security strength provided by an algorithm or key may be reduced or lost completely. For example, the algorithm or key length used may no longer offer adequate security because of improvements in computational capability or cryptanalysis. In this case, applying protection to "new" information can be performed using stronger algorithms or keys. However, information that was previously protected using these algorithms and keys may no longer be secure. This information may include other keys, or other sensitive data protected by the keys. A reduction in the security strength provided by an algorithm or key has the following implications:

- Encrypted information: The security of encrypted information that was exposed at any time to unauthorized entities in its encrypted form should be considered suspect. For example, keys that were transmitted in encrypted form using an algorithm or key length that is later broken (e.g., a key-wrapping key or key-transport key), may need to be considered as compromised, since an adversary could have saved the encrypted form of the keys for later decryption if methods for breaking the algorithm are eventually found (see Section 5.5 for a discussion of key compromise). Even if the transmitted, encrypted information is subsequently re-encrypted for storage using a different key or algorithm, the information may already be compromised because of the weakness of the transmission algorithm or key. However, encrypted information that was not exposed in this manner (e.g., not transmitted) may still be secure, even though the encryption algorithm or key length no longer provides adequate protection. For example, if the encrypted form of the keys and the information protected by those keys was never transmitted, then the information may still be confidential. The lessons to be learned are that an encryption mechanism used for information that will be exposed to unauthorized entities (e.g., via transmission) should provide a high level of security protection, and the use of each key should be limited (i.e., the cryptoperiod should be short) so that a compromised key cannot be used to reveal very much information. If the algorithm itself is broken, an adversary is forced to perform more work when each key is used to encrypt a very limited amount of information. See Section 5.3.6 for a discussion about cryptoperiods.

- Digital signatures on stored data[22]: Digital signatures may be computed on data prior to transmission and subsequent storage. In this case, both the signed data and the digital signature would be stored. If the security strength of the signature is later reduced (e.g., because of a break of the algorithm), the signature may still be valid if the stored data and its associated digital signature have been adequately protected from modification since a time prior to the reduction in strength (e.g., by applying a digital signature using a stronger algorithm or key). See Section 5.5, item 1 for further discussion. Storage capabilities are being developed that employ cryptographic timestamps to store digitally signed data beyond the normal security life of the original signature mechanism or its keys.

- Symmetric authentication codes on stored data[23]: Symmetric authentication codes may be computed on data prior to transmission and subsequent storage. If the received data and authentication code are stored as received, and the security strength of the authentication algorithm or key is later reduced (e.g., because of a break of the algorithm), the authentication code may still be valid if the stored data and its associated authentication code have been adequately protected from modification since a time prior to the reduction in strength (e.g., by applying another authentication code using a stronger algorithm or key). See Section 5.5, item 1 for further discussion. Storage capabilities are being developed that employ cryptographic timestamps to store authenticated data beyond the normal security life of the original authentication mechanism or its keys.

---

[22] Digital signatures on data that is not stored are not considered, as their value is considered to be short-lived, e.g., the digital signature was intended to be used to detect errors introduced during transmission only.

[23] Symmetric authentication codes on data that is not stored, as their value is considered to be short-lived.

6   Protection Requirements for Cryptographic Information

This section gives guidance on the types of protection required for keying material. Cryptographic keying material is defined as the cryptographic key and associated information required to use the key. The specific information varies, depending on the type of key. The cryptographic keying material must be protected in order for the security services to be "meaningful." A FIPS140-validated cryptographic module may provide much of the protection needed; however, whenever the keying material exists external to a [FIPS140] cryptographic module, additional protection is required. The type of protection needed depends on the type of key and the security service for which the key is used.

6.1   Protection and Assurance Requirements

Keying material **should** be (operationally) available as long as the associated cryptographic service is required. Keys may be maintained within a cryptographic module while they are being actively used, or they may be stored externally (provided that proper protection is afforded) and recalled as needed. Some keys may need to be archived if required beyond the key's originator-usage period (see Section 5.3.5).

The following protections and assurances may be required for the keying material.

Integrity protection (also called assurance of integrity) **shall** be provided for all keying material. Integrity protection always involves checking the source and format of received keying material (see Section 5.4.1). Integrity protection can be provided by cryptographic integrity mechanisms (e.g. cryptographic checksums, cryptographic hash functions, MACs, and signatures), non-cryptographic integrity mechanisms (e.g. CRCs, parity checks, etc.) (see Appendix A), or physical protection mechanisms. Guidance for the selection of appropriate integrity mechanisms is given in Sections 6.2.1.2 and 6.2.2.2.

Confidentiality protection for all symmetric and private keys **shall** be provided. Public keys generally do not require confidentiality protection. When the symmetric or private key exists internal to a validated cryptographic module, confidentiality protection is provided by the cryptographic module in accordance with [FIPS140], level 2 or higher. When the symmetric or private key exists external to the cryptographic module, confidentiality protection **shall** be provided either by encryption (e.g., key wrapping) or by controlling access to the key via physical means (e.g. storing the keying material in a safe with limited access). The security and operational impact of specific confidentiality mechanisms varies. Guidance for the selection of appropriate confidentiality mechanisms is given in Sections 6.2.1.3 and 6.2.2.3.

Association protection **shall** be provided for a cryptographic security service by ensuring that the correct keying material is used with the correct data in the correct application or equipment. Guidance for the selection of appropriate association protection is given in Sections 6.2.1.4 and 6.2.2.4.

Assurance of domain-parameter and public-key validity provides confidence that the parameters and keys are arithmetically correct (see Sections 5.4.2 and 5.4.3). Guidance for the selection of appropriate validation mechanisms is given in [SP800-56A] and [SP800-89], as well as this document.

Assurance of private key possession provides assurance that the owner of a public key actually possesses the corresponding private key (see Section 5.4.4).

The *period of protection* for cryptographic keys, associated key information, and cryptographic parameters (e.g. initialization vectors) depends on the type of key, the associated cryptographic service, and the length of time for which the cryptographic service is required. The period of protection includes the cryptoperiod of the key (see Section 5.3). The period of protection is not necessarily the same for integrity as it is for confidentiality. Integrity protection may be required until a key is no longer used, but confidentiality protection may be required until the key is destroyed.

6.1.1 Summary of Protection and Assurance Requirements for Cryptographic Keys

Table 5 provides a summary of the protection requirements for keys during distribution and storage. Methods for providing the necessary protection are discussed in Section 6.2.

Guide to Table 5:

a. Column 1 (Key Type) identifies the key types.

b. Column 2 (Security Service) indicates the type of security service that is provided by the key in conjunction with a cryptographic technique.

c. Column 3 (Security Protection) indicates the type of protection required for the key (i.e., integrity, and confidentiality).

d. Column 4 (Association Protection) indicates the types of associations that need to be protected for that key, such as associating the key with the usage or application, the authorized communications participants or other indicated information. The association with domain parameters applies only to algorithms where they are used.

e. Column 5 (Assurances Required) indicates whether assurance of public-key validity and/or assurance of private-key possession needs to be obtained as defined in [SP800-56A], [SP800-56B], [SP800-89] and this Recommendation. Assurance of public-key validity provides a degree of confidence that a key is arithmetically correct. See Section 5.4.3 for further details. Assurance of private-key possession provides a degree of confidence that the entity providing a public key actually possessed the associated private key at some time. See Section 5.4.4 for further details.

f. Column 6 (Period of Protection) indicates the length of time that the integrity and/or confidentiality of the key need to be maintained (see Section 5.3). Symmetric keys and private keys **shall be** destroyed at the end of their period of protection (see Sections 8.3.4 and 9.3).

Table 5: Protection requirements for cryptographic keys

Key Type	Security Service	Security Protection	Association Protection	Assurances Required	Period of Protection
Private signature key	Authentication; Integrity; Non-repudiation	Integrity[24]; Confidentiality	Usage or application; Domain parameters; Public signature-verification key	Possession	From generation until the end of the cryptoperiod

---

[24] Integrity protection can be provided by a variety of means. See Sections 6.2.1.2 and 6.2.2.2.

Key Type	Security Service	Security Protection	Association Protection	Assurances Required	Period of Protection
Public signature-verification key	Authentication; Integrity; Non-repudiation	Integrity;	Usage or application; Key pair owner; Domain parameters; Private signature key; Signed data	Validity	From generation until no protected data needs to be verified
Symmetric authentication key	Authentication; Integrity	Integrity; Confidentiality	Usage or application; Other authorized entities; Authenticated data		From generation until no protected data needs to be verified
Private authentication key	Authentication; Integrity	Integrity; Confidentiality	Usage or application; Public authentication key; Domain parameters	Possession	From generation until the end of the cryptoperiod
Public authentication key	Authentication; Integrity	Integrity	Usage or application; Key pair owner; Authenticated data; Private authentication key; Domain parameters	Validity	From generation until no protected data needs to be authenticated
Symmetric data-encryption/ decryption key	Confidentiality	Integrity; Confidentiality	Usage or application; Other authorized entities; Plaintext/Encrypted data		From generation until the end of the lifetime of the data or the end of the cryptoperiod, whichever comes later
Symmetric key-wrapping key	Support	Integrity; Confidentiality	Usage or application; Other authorized entities; Encrypted keys		From generation until the end of the cryptoperiod or until no wrapped keys require protection, whichever is later.
Symmetric and asymmetric RNG keys	Support	Integrity; Confidentiality	Usage or application	Possession of private RNG key, if used	From generation until replaced

75

Key Type	Security Service	Security Protection	Association Protection	Assurances Required	Period of Protection
Symmetric master key	Support	Integrity; Confidentiality	Usage or application; Other authorized entities; Derived keys		From generation until the end of the cryptoperiod or the end of the lifetime of the derived keys, whichever is later.
Private key-transport key	Support	Integrity; Confidentiality	Usage or application; Encrypted keys; Public key-transport key	Possession	From generation until the end of the period of protection for all transported keys
Public key-transport key	Support	Integrity	Usage or application; Key pair owner; Private key-transport key	Validity	From generation until the end of the cryptoperiod
Symmetric key-agreement key	Support	Integrity; Confidentiality	Usage or application; Other authorized entities		From generation until the end of the cryptoperiod or until no longer needed to determine a key, whichever is later
Private static key-agreement key	Support	Integrity; Confidentiality	Usage or application; Domain parameters; Public static key-agreement key	Possession	From generation until the end of the cryptoperiod or until no longer needed to determine a key, whichever is later
Public static key-agreement key	Support	Integrity	Usage or application; Key pair owner; Domain parameters; Private static key-agreement key	Validity	From generation until the end of the cryptoperiod or until no longer needed to determine a key, whichever is later
Private ephemeral key-agreement key	Support	Integrity; Confidentiality	Usage or application; Public ephemeral key-agreement key; Domain parameters;		From generation until the end of the key-agreement process. After the end of the process, the key **shall** be destroyed

Key Type	Security Service	Security Protection	Association Protection	Assurances Required	Period of Protection
Public ephemeral key-agreement key	Support	Integrity[25]	Key pair owner; Private ephemeral key-agreement key; Usage or application; Domain parameters	Validity	From generation until the key-agreement process is complete
Symmetric authorization keys	Authorization	Integrity; Confidentiality	Usage or application; Other authorized entities		From generation until the end of the cryptoperiod of the key
Private authorization key	Authorization	Integrity; Confidentiality	Usage or application; Public authorization key; Domain parameters	Possession	From generation until the end of the cryptoperiod of the key
Public authorization key	Authorization	Integrity	Usage or application; Key pair owner; Private authorization key; Domain parameters	Validity	From generation until the end of the cryptoperiod of the key

6.1.2   Summary of Protection Requirements for Other Cryptographic or Related Information

Table 6 provides a summary of the protection requirements for other cryptographic information during distribution and storage. Mechanisms for providing the necessary protection are discussed in Section 6.2.

---

[25] Public ephemeral key-agreement keys are not generally protected during transmission; however, the key-agreement protocols may be designed to detect unauthorized substitutions and modifications to the transmitted ephemeral public keys. In this case, the protocols form the data integrity mechanism.

Guide to Table 6:

a. Column 1 (Cryptographic Information Type) identifies the type of cryptographic information.

b. Column 2 (Security Service) indicates the type of security service provided by the cryptographic information.

c. Column 3 (Security Protection) indicates the type of security protection for the cryptographic information.

d. Column 4 (Association Protection) indicates the relevant types of associations for each type of cryptographic information.

e. Column 5 (Assurance of Domain Parameter Validity) indicates the cryptographic information for which assurance **shall** be obtained as defined in [SP800-56A] and [SP800-89] and in Section 5.4 of this Recommendation. Assurance of domain-parameter validity gives confidence that domain parameters are arithmetically correct.

f. Column 6 (Period of Protection) indicates the length of time that the integrity and/or confidentiality of the cryptographic information needs to be maintained. The cryptographic information **shall** be destroyed at the end of the period of protection (see Sections 8.3.4).

Table 6: Protection requirements for other cryptographic or related material

Crypto. Information Type	Security Service	Security Protection	Association Protection	Assurance of Domain Parameter Validity	Period of Protection
Domain parameters	Depends on key assoc. with the parameters	Integrity	Usage or application; Private and public keys	Yes	From generation until no longer needed to generate keys or verify signatures
Initialization vectors	Depends on algorithm	Integrity[26]	Protected data		From generation until no longer needed to process the protected data
Shared secrets	Support	Confidentiality; Integrity			From generation until the end of the transaction. The shared secret **shall** be destroyed at the end of the period of protection
RNG Seeds	Support	Confidentiality; Integrity	Usage or application		Used once and destroyed immediately after use

[26] IVs are not generally protected during transmission; however, the decryption system may be designed to detect or minimize the effect of unauthorized substitutions and modifications to transmitted IVs. In this case the decryption system is the data-integrity mechanism.

Crypto. Information Type	Security Service	Security Protection	Association Protection	Assurance of Domain Parameter Validity	Period of Protection
Other public information	Support	Integrity	Usage or application; Other authorized entities; Data processed using the nonce		From generation until no longer needed to process data using the public information
Other secret information	Support	Confidentiality; Integrity	Usage or application; Other authorized entities; Data processed using the secret information		From generation until no longer needed to process data using the secret information
Intermediate results	Support	Confidentiality; Integrity	Usage or application		From generation until no longer needed and the intermediate results are destroyed
Key-control information (e.g., IDs, purpose)	Support	Integrity	Key		From generation until the associated key is destroyed
Random number	Support	Integrity; Confidentiality (depends on usage)			From generation until no longer needed, and the random number is destroyed
Password	Authentication; Key derivation	Integrity; Confidentiality	Usage or application; Owning entity		From generation until replaced or no longer needed to authenticate the entity or to derive keys
Audit information	Support	Integrity; Access authorization	Audited events; Key control information		From generation until no longer needed

6.2 Protection Mechanisms

During the lifetime of cryptographic information, the information is either "in transit" (e.g., is in the process of being manually distributed or distributed using automated protocols to the authorized communications participants for use by those entities) or is "at rest" (e.g., the information is in storage). In either case, the keying material **shall** be protected in accordance with Section 6.1. However, the choice of protection mechanisms may vary. Although several methods of protection are provided in the following subsections, not all methods provide equal security. The method **should** be carefully selected. In addition, the mechanisms prescribed do not, by themselves, guarantee protection. The implementation and the associated key management need to provide adequate security to prevent any feasible attack from being successful.

6.2.1 Protection Mechanisms for Cryptographic Information in Transit

Cryptographic information in transit may be keying material being distributed in order to obtain a cryptographic service (e.g., establish a key that will be used to provide confidentiality) (see Section 8.1.5), or cryptographic information being backed up or archived for possible use or recovery in the future (see Sections 8.2.2 and 8.3.1). This may be accomplished manually (i.e., via a trusted courier), in an automated fashion (i.e., using automated communication protocols) or by some combination of manual and automated methods. For some protocols, the protections are provided by the protocol; in other cases, the protection for the keying material is provided directly on the keying material. It is the responsibility of the originating entity to apply protection mechanisms, and the responsibility of the recipient to undo or check the mechanisms used.

6.2.1.1 Availability

Since communications may be garbled, intentionally altered, or destroyed, the availability of cryptographic information after transit cannot be assured using cryptographic methods. However, availability can be supported by redundant or multiple channels, store and forward systems (deleting by the sender only after confirmation of receipt), error correction codes, and other non-cryptographic mechanisms.

Communication systems **should** incorporate non-cryptographic mechanisms to ensure the availability of transmitted cryptographic information after it has been successfully received, rather than relying on retransmission by the original sender for future availability

6.2.1.2 Integrity

Integrity protection involves both the prevention and detection of modifications to information. When modifications are detected, measures may be taken to restore the information to its unaltered form. Cryptographic mechanisms are often used to detect unauthorized modifications. The integrity of cryptographic information during transit **shall** be protected using one or more of the following mechanisms:

1. Manual method (physical protection is provided):

   (a) An integrity mechanism comparable to a CRC (e.g., CRC, MAC or digital signature) is used on the information, and the resulting code (e.g., CRC, MAC or digital signature) is provided to the recipient. Note: A CRC may be used instead of a MAC or digital signature, since the physical protection is intended to protect against intentional modifications.

   -OR-

   (b) The keying material is used to perform the intended cryptographic operation. If the received information does not conform to the expected format, or the data is inconsistent in the context of the application, then the keying material may have been corrupted.

2. Automated distribution via communication protocols (provided by the user or by the communication protocol):

   (a) An **approved** cryptographic integrity mechanism (e.g., a MAC or digital signature) is used on the information, and the resulting code (e.g., a MAC or digital signature) is provided to the recipient. Note that a CRC is not **approved** for this purpose. The integrity mechanism may be applied only to the cryptographic information, or may be applied to an entire message

-OR-

(b) The keying material is used to perform the intended cryptographic operation. If the use of the keying material produces incorrect results, or the data is inconsistent in the context of the application, then the received keying material may have been corrupted.

The response to the detection of an integrity failure will vary, depending on the specific environment. Improper error handling can allow attacks (e.g., side channel attacks). A security policy (see Part 2) **should** define the response to such an event. For example, if an error is detected in the received information, and the receiver requires that the information is entirely correct (e.g., the receiver cannot proceed when the information is in error), then:

a. The information **should not** be used,

b. The recipient may request that the information be resent (retransmissions **should** be limited to a predetermined number of times), and

c. Information related to the incident may be stored in an audit log to later identify the source of the error.

6.2.1.3   Confidentiality

Keying material may require confidentiality protection during transit. If confidentiality protection is required, the keying material **shall** be protected using one or more of the following mechanisms:

1. Manual method:

(a) The keying material is encrypted

-OR-

(b) The keying material is separated into key components. Each key component is handled, using split knowledge procedures (see Sections 8.1.5.2.1 and 8.1.5.2.2.1), so that no single individual can acquire access to all key components.

-OR-

(c) Appropriate physical and procedural protection is provided (e.g., by using a trusted courier).

2. Automated distribution via communication protocols: The keying material is encrypted using an **approved** algorithm and key size.

6.2.1.4   Association with Usage or Application

The association of keying material with its usage or application **shall** be either specifically identified during the distribution process or be implicitly defined by the use of the application. See Section 6.2.3 for for a discussion of the metadata associated with keys.

6.2.1.5   Association with Other Entities

The association of keying material with the appropriate entity (e.g., the key source) **shall** be either specifically identified during the distribution process (e.g., using public-key certificates) or be implicitly defined by the use of the application. See Section 6.2.3 for a discussion of the metadata associated with keys.

6.2.1.6   Association with Other Related Information

Any association with other related information (e.g., domain parameters, the encryption/decryption key or IVs) **shall** be either specifically identified during the distribution process or be implicitly defined by the use of the application. See Section 6.2.3 for a discussion of the metadata associated with the other related information.

6.2.2   Protection Mechanisms for Information in Storage

Cryptographic information that is not in transit is at rest in some device or storage media. This may include copies of the information that is also in transit. Information-at-rest (i.e., stored information) **shall** be protected in accordance with Section 6.1. A variety of protection mechanisms may be used.

The cryptographic information may be stored so as to be immediately available to an application (e.g., on a local hard disk or a server); this would be typical for keying material stored within a cryptographic module or in immediately accessible storage (e.g., on a local hard drive). The keying material may also be stored in electronic form on a removable media (e.g., a CD-ROM), in a remotely accessible location, or in hard copy form and placed in a safe; this would be typical for backup or archive storage.

6.2.2.1   Availability

Cryptographic information may need to be readily available for as long as data is protected by the information. A common method for providing this protection is to make one or more copies of the cryptographic information and store them in separate locations. During a key's cryptoperiod, keying material requiring long-term availability **should** be stored in both normal operational storage (see Section 8.2.1) and in backup storage (see Section 8.2.2.1). Cryptographic information that is retained after the end of a key's cryptoperiod **should** be placed in archive storage (see Section 8.3.1). This Recommendation does not preclude the use of the same storage media for both backup and archive storage.

Specifics on the long-term availability requirement for each key type are addressed for backup storage in Section 8.2.2.1, and for archive storage in Section 8.3.1.

The recovery of this cryptographic information for use in replacing cryptographic information that is lost (e.g., from normal storage), or in performing cryptographic operations after the end of a key's cryptoperiod is discussed in Sections 8.2.2.2 (recovery during normal operations) and 8.3.1 (recovery from archive storage), and in Appendix B.

6.2.2.2   Integrity

Integrity protection is concerned with ensuring that the information is correct. Absolute protection against modification is not possible. The best that can be done is to use reasonable measures to prevent modifications, to use methods to detect any modifications that occur (with a very high probability), and to restore the information to its original content when modifications have been detected.

All cryptographic information requires integrity protection. Integrity protection **shall** be provided by physical mechanisms, cryptographic mechanisms or both.

Physical mechanisms include:

1. A validated cryptographic module or operating system that limits access to the stored information,

2. A computer system or media that is not connected to other systems,

3. A physically secure environment with appropriate access controls that is outside a computer system (e.g., in a safe with limited access).

Cryptographic mechanisms include:

a. An **approved** cryptographic integrity mechanism (e.g., a MAC or digital signature) that is computed on the information and is later used to verify the integrity of the stored information.

b. Performing the intended cryptographic operation. If the received information is incorrect, it is possible that the keying material may have been corrupted.

In order to restore the cryptographic information when an error is detected, multiple copies of the information **should** be maintained in physically separate locations (i.e., in backup or archive storage; see Sections 8.2.2.1 and 8.3.1). The integrity of each copy **should** be periodically checked.

6.2.2.3   Confidentiality

One of the following mechanisms **shall** be used to provide confidentiality for private or secret keying material in storage:

1. Encryption with an **approved** algorithm in a [FIPS140] cryptographic module. It **shall** be no easier to recover the key-encrypting key than it is to recover the key being encrypted,

    -OR-

2. Physical protection provided by a [FIPS140] (level 2 or higher) cryptographic module,

    -OR-

3. Physical protection provided by secure storage with controlled access (e.g., a safe or protected area).

6.2.2.4   Association with Usage or Application

Cryptographic information is used with a given cryptographic mechanism (e.g., digital signatures or key establishment) or with a particular application. Protection **shall** be provided to ensure that the information is not used incorrectly (e.g., not only must the usage or application be associated with the keying material, but the integrity of this association must be maintained). This protection can be provided by separating the cryptographic information from that of other mechanisms or applications, or by the use of appropriate metadata associated with the information. Section 6.2.3 addresses the metadata associated with cryptographic information.

6.2.2.5   Association with the Other Entities

Some cryptographic information needs to be correctly associated with another entity (e.g., the key source), and the integrity of this association **shall** be maintained. For example, a symmetric (secret) key used for the encryption of information, or the computation of a MAC needs to be associated with the other entity(ies) that shares the key. Public keys need to be correctly associated (e.g., cryptographically bound) with the owner of the key pair (e.g., using public-key certificates).

The cryptographic information **shall** retain its association during storage by separating the information by "entity" or application, or by using appropriate metadata for the information. Section 6.2.3 addresses the metadata used for cryptographic information.

6.2.2.6   Association with Other Related Information

An association may need to be maintained between protected information and the keying material that protected that information. In addition, keys may require association with other keying material (see Section 6.2.1.6).

Storing the information together or providing some linkage or pointer between the information accomplishes the association. Often, the linkage between a key and the information it protects is accomplished by providing an identifier for a key, storing the identifier with the key in the key's metadata, and storing the key's identifier with the protected information. The association **shall** be maintained for as long as the protected information needs to be processed.

Section 6.2.3 addresses the use of metadata for cryptographic information.

6.2.3   Metadata Associated with Cryptographic Information

Metadata may be used with cryptographic information to define the use of that information or to provide a linkage between cryptographic information.

6.2.3.1   Metadata for Keys

Metadata is used to identify attributes, parameters, or the intended use of a key, and as such contains the key's control information. Different applications may require different metadata elements for the same key type, and different metadata elements may be required for different key types. It is the responsibility of an implementer to select suitable metadata elements for keys. When metadata is used, the metadata **should** accompany a key (i.e., the metadata is typically stored or transmitted with a key). Some examples of metadata elements are:

1. Key identifier
2. Information identifying associated keys (e.g., the association between a public and private key)
3. Identity of the key's owner or the sharing entity(ies)
4. Cryptoperiod (e.g., start date and end date)
5. Key type (e.g., signing private key, encryption key, master key)
6. Application (e.g., purchasing, email)

7. Sensitivity of the information protected by the key

8. Counter[27]

9. Domain parameters (e.g., the domain parameters used by DSA or ECDSA, or a pointer to them)

10. Status or state of the key

11. Key-encrypting key identifier (e.g., key-wrapping key identifier, algorithm for the key-wrapping algorithm, etc.)

12. Integrity-protection mechanism (e.g., key and algorithm used to provide cryptographic protection, and the protection code (e.g., MAC, digital signature))

13. Other information (e.g., the length of the key, any protection requirements, who has access rights to the key, additional conditions for use)

6.2.3.2   Metadata for Related Cryptographic Information

Cryptographic information other than keying material may need metadata to "point to" the keying material that was used to provide the cryptographic protection for the information. The metadata may also contain other related cryptographic information. When metadata is used, the metadata **should** accompany the information (i.e., the metadata is typically stored or transmitted with the information) and contain some subset of the following information:

1. The type of information (e.g., domain parameters)

2. Source of the information (e.g., the entity that sent the information)

3. Application (e.g., purchasing, email)

4. Other associated cryptographic information (e.g., a key, MAC or hash value)

5. Any other information (e.g., who has access rights).

---

[27] Used to detect the playback of a previously-transmitted key package

7 Key States and Transitions

A key may pass through several states between its generation and its destruction.

7.1 Key States

A key is used differently, depending upon its state in the key's lifecycle. Key states are defined from a system point-of-view, as opposed to the point-of-view of a single cryptographic module. The following is a list of the states that a key may assume; additional states may be applicable for some systems.

1. **Pre-activation state**: The key has been generated, but has not been authorized for use. In this state, the key may only be used to perform proof-of-possession or key confirmation. Other than for proof-of-possession (Section 8.1.5.1.1.2) or key-confirmation (Section 4.2.5.5) purposes, a key **shall not** be used to apply cryptographic protection to information (e.g., encrypt or sign information to be transmitted or stored) or to process cryptographically protected information (e.g., decrypt ciphertext or verify a digital signature) while in this state.

2. **Active state**: The key may be used to cryptographically *protect* information (e.g., encrypt plaintext or generate a digital signature), to cryptographically *process* previously protected information (e.g., decrypt ciphertext or verify a digital signature) or both. When a key is active, it may be designated for protection only, processing only, or both protection and processing. For example, private signature keys and public key-transport keys are implicitly designated for protection only; public signature-verification keys and private key-transport keys are designated for processing only. A symmetric data-encryption key may be used to encrypt data during its originator-usage period and decrypt the encrypted data during its recipient-usage period (see Section 5.3.5); at the end of its cryptoperiod, the symmetric key **shall** transition to the deactivated state.

3. **Deactivated state**: A key whose cryptoperiod has expired but may still needed to perform cryptographic processing is deactivated until it is destroyed. A deactivated key **shall not** be used to apply cryptographic protection to information, but in some cases, it may be used to process cryptographically protected information. When a key in the deactivated state is no longer required for processing cryptographically protected information, the key **should** be destroyed (see Section 8.3.4).

4. **Destroyed state**: The key has been destroyed as specified in Section 8.3.4. Even though the key no longer exists in this state, certain key attributes (e.g., key name, type, and cryptoperiod) may have been retained (see Section 8.4).

5. **Compromised state**: Generally, keys are compromised when they are released to or determined by an unauthorized entity. If the integrity or secrecy of the key is suspect, the key **shall not** be used to apply cryptographic protection to information; in some cases, a compromised key may be used to process cryptographically protected information, even though the confidentiality, integrity, non-repudiation or associations of the information may be suspect. For example, a signature may be validated if it can be shown that the signed data with its signature has been physically protected since a time before the compromise occurred. This processing **shall** be done only under very highly controlled

conditions, where the users of the information are fully aware of the possible consequences.

A compromised key **shall** be revoked (see Section 9.3.4). The compromised state may be entered from all states except the destroyed and destroyed compromised states.

6. **Destroyed Compromised state**: The key has been destroyed after a compromise, or the key has been destroyed, and a compromise has later been discovered. Key attributes (e.g., key name, type, and cryptoperiod) may have been retained. This state differs from the destroyed state in that keys in this state are known or suspected of being compromised (see Section 8.4).

7.2   Key State Transitions

Transitions between states are triggered by events, such as the expiration of a cryptoperiod or the detection of a compromise of a key. Figure 3 depicts the key states from Section 7.1 and the transitions between the states.

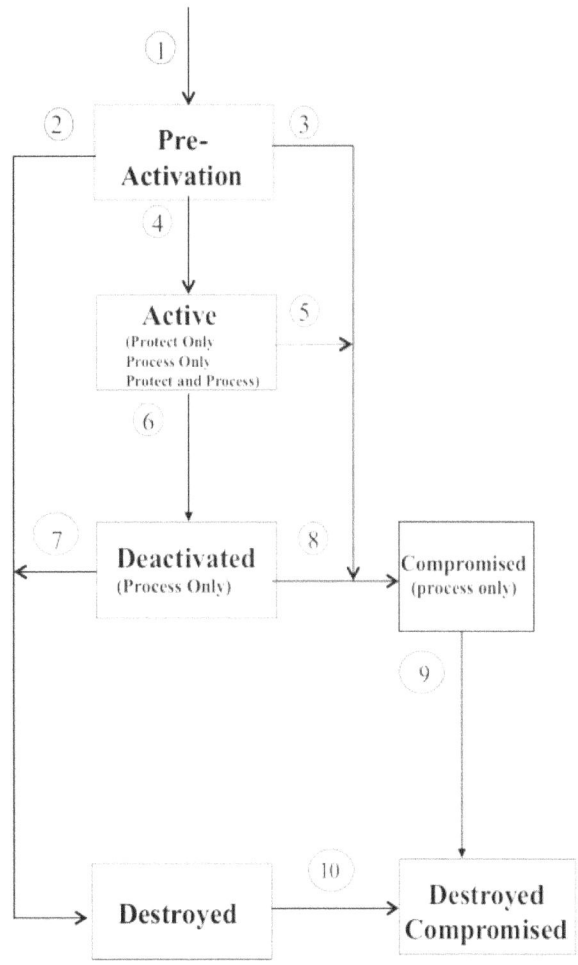

Figure 3: Key states and transitions

Transition 1:   A key enters the pre-activation state immediately upon generation.

Transition 2: A key that is never used **should** transition from the pre-activation state directly to the destroyed state. In this case, the integrity of a key or the confidentiality of a key requiring confidentiality protection is considered trustworthy, but it has been determined that the key will not be needed in the future.

Transition 3: A key that is never used **shall** transition from the pre-activation state to the compromised state when the integrity of a key or the confidentiality of a key requiring confidentiality protection becomes suspect before first use.

Transition 4: Keys **shall** transition from the pre-activation state to the active state when the key becomes available for use. This transition may occur upon reaching an activation date or may occur because of an external event. In the case where keys are generated for immediate use, this transition occurs immediately after entering the pre-activation state.

This transition marks the beginning of a key's cryptoperiod (see Section 5.3).

Transition 5: An active key **shall** transition from the active state to the compromised state when the integrity of a key or the confidentiality of a key requiring confidentiality protection becomes suspect. Generally, keys are compromised when they are released to or determined by an unauthorized entity.

Transition 6: An active key **shall** transition to the deactivated state if it is no longer to be used to apply cryptographic protection to data and no longer intended to be used to process cryptographically protected data. A key **shall** transition from the active state to the deactivated state as a result of a revocation action (see Section 8.3.5) for a reason other than a key compromise, or if the key is replaced (see Section 8.2.3), or at the end of the key's cryptoperiod (see Sections 5.3.4 and 5.3.5).

Transition 7: Assuming that a key is not determined to be compromised while in the deactivated state, a key **should** transition from the deactivated state to the destroyed state as soon as it is no longer needed

Transition 8: A deactivated key **shall** transition from the deactivated state to the compromised state when the integrity of a key or the confidentiality of a key requiring confidentiality protection becomes suspect. Generally, keys are compromised when they are released to or determined by an unauthorized entity.

Transition 9: A key in the compromised state **should** transition to the destroyed compromised state when the key is no longer needed to process protected data.

Transition 10: A destroyed key **should** transition to the destroyed compromised state if it is determined that the key was previously compromised. Although the key itself has already been destroyed, transition to the destroyed compromised state **should** be indicated in any remaining key attributes for that key.

7.3 States and Transitions for Asymmetric Keys

The preceding discussion of key states and transitions applies to both symmetric and asymmetric keys; however, some observations that are specific to asymmetric keys are in order.

Asymmetric keys that are or will be certified **shall** be in the pre-activation state until certified or until the "not before" date specified in a certificate has passed. The types of transitions for asymmetric keys depend on the key type. Examples of transitions follow:

a. A private signature key **shall not** be retained in the deactivated state, but transition immediately to the destroyed state.

b. A private signature key transitioning from the active state to the compromised state **shall not** be retained in that state, but transition immediately to the destroyed-compromised state unless retention is required for legal purposes.

c. A public signature-verification key **shall** transition to the deactivated state at the end of the corresponding private key's cryptoperiod. The public signature-verification key **shall** enter the compromised state if its integrity, or the confidentiality or integrity of its corresponding private signature key become suspect. However, public signature-verification keys need not be destroyed.

d. A private key-transport key transitions from the active state to the deactivated state when its corresponding public key is no longer to be used to apply cryptographic protection. The private key-transport key **shall** enter the compromised state if its integrity or confidentiality become suspect.

e. A public key-transport key transitioning from the active state to the deactivated state may be retained in that state, or transition to the destroyed state. The key **shall** enter the compromised state when its integrity, or the confidentiality or integrity of its corresponding private key-transport key is suspect.

f. Static private and public key-agreement keys, and ephemeral public key-agreement keys transitioning from the active state to the deactivated state may be retained in that state, or transition to the destroyed state.

g. Ephemeral private key-agreement keys **shall** be destroyed immediately after use (see SP800-56A]). That is, immediately after the use of an ephemeral private key-agreement key, the key transitions through the deactivated state to the destroyed state.

8   Key-Management Phases and Functions

The cryptographic key-management lifecycle can be divided into four phases. During each phase, the keys are in certain specific states as discussed in Section 7. In addition, within each phase, certain key-management functions are typically performed. These functions are necessary for the management of the keys and their associated attributes.

Key-management information is characterized by attributes. The attributes required for key management might include the identity of a person or system associated with that key or the types of information that person is authorized to access. Attributes are leveraged by applications to select the appropriate cryptographic key(s) for a particular service. While these attributes do not appear in cryptographic algorithms, they are crucial to the implementation of applications and application protocols.

The four phases of key management are specified below.

1. **Pre-operational phase:** The keying material is not yet available for normal cryptographic operations. Keys may not yet be generated, or are in the pre-activation state. System or enterprise attributes are established during this phase as well.

2. **Operational phase:** The keying material is available and in normal use. Keys are in the active state. Keys may be designated as protect only, process only, or protect and process.

3. **Post-operational phase**: The keying material is no longer in normal use, but access to the keying material is possible, and the keying material may be used for process only in certain circumstances. Keys are in the deactivated or compromised states. Keys in the post-operational phase are archived (see section 8.3.1) when not processing data.

4. **Destroyed phase:** Keys are no longer available. All records of their existence may have been deleted. Keys are in the destroyed or destroyed compromised states. Although the keys themselves are destroyed, the key attributes (e.g., key name, type, cryptoperiod, and usage period) may be retained (see Section 8.4).

A flow diagram for the key management phases is presented in Figure 4. Seven phase transitions are identified in the diagram. A key **shall not** be able to transfer back to any previous phase.

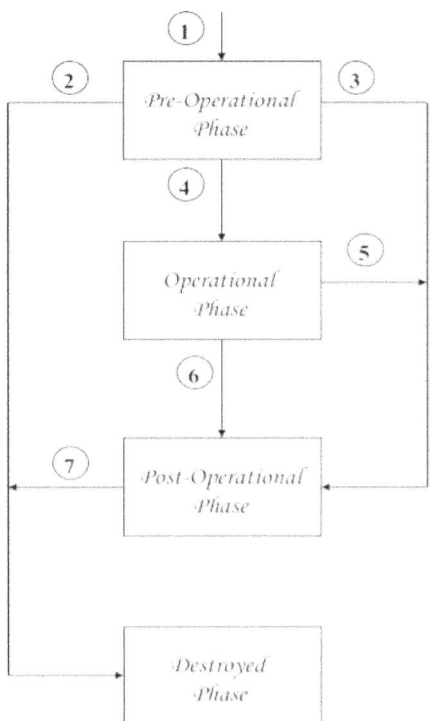

Figure 4: Key management phases

Transition 1: A key is in the pre-operational phase upon generation (pre-activation state).

Transition 2: If keys are produced, but never used, they may be destroyed by transitioning from the pre-operational phase directly to the destroyed phase.

Transition 3: When a key in the pre-operational phase is compromised, it transitions to the post-operational phase (compromised state).

Transition 4: After the required key attributes have been established, keying material has been generated, and the attributes are associated with the key during the pre-operational phase, the key is ready to be used by applications and transitions to the operational phase at the appropriate time.

Transition 5: When a key in the operational phase is compromised, it transitions to the post-operational phase (compromised state).

Transition 6: When keys are no longer required for normal use (i.e., the end of the cryptoperiod has been reached and the key is no longer "active"), but access to those keys needs to be maintained, the key transitions to the post-operational phase.

Transition 7: Some applications will require that access be preserved for a period of time, and then the keying material may be destroyed. When it is clear that a key in the post-operational phase is no longer needed, it may transition to the destroyed phase.

The combination of key states and key phases is illustrated in Figure 5. The pre-operational and operational phases contain only one state each, while the post-operational and destroyed phases have two states each.

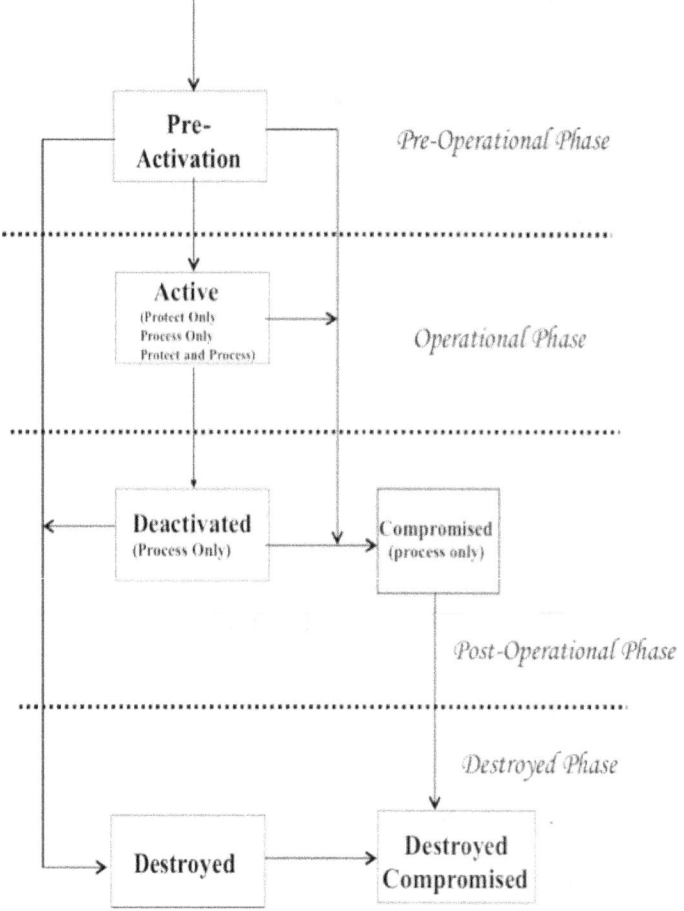

Figure 5: Key management states and phases

The following subsections discuss the functions that are performed in each phase of key management. A key-management system may not have all identified functions, since some functions may not be appropriate. In some cases, one or more functions may be combined, or the functions may be performed in a different order. For example, a system may omit the functions of the post-operational phase if keys are never archived, and compromised keys are immediately destroyed. In this case, keys would move from the operational phase directly to the destroyed phase.

8.1   Pre-operational Phase

During the pre-operational phase of key management, keying material is not yet available for normal cryptographic operations.

8.1.1   User Registration Function

During user registration, an entity interacts with a registration authority to become an authorized member of a security domain. In this phase, a user identifier or device name may be established to identify the member during future transactions. In particular, security infrastructures may associate the identification information with the entity's keys (see Sections 8.1.5 and 8.1.6). The entity may also establish various attributes during the registration function, such as email addresses or role/authorization information. As with identity information, these attributes may be associated with the entity's keys by the infrastructure to support secure application-level security services.

Since applications will depend upon the identity established during this process, it is crucial that the registration authority establish appropriate procedures for the validation of identity. Identity may be established through an in-person appearance at a registration authority, or may be established entirely out-of-band. Human entities are usually required to provide credentials (e.g., an identification card or birth certificate), while system entities are vouched for by those individuals responsible for system operation. The strength (or weakness) of a security infrastructure will often depend upon the identification process.

User and key registration (see Section 8.1.6) may be performed separately, or in concert. If performed separately, the user registration process will generally establish a secret value (e.g., a password, PIN, or HMAC key); the secret value may be used to authenticate the user during the key registration step. If performed in concert, the user establishes an identity and performs key registration in the same process, so the secret value is not required.

8.1.2   System Initialization Function

System initialization involves setting up or configuring a system for secure operation. This may include algorithm preferences, the identification of trusted parties, and the definition of domain-parameter policies and any trusted parameters (e.g., recognized certificate policies).

8.1.3   User Initialization Function

User initialization consists of an entity initializing its cryptographic application (e.g., installing and initializing software or hardware). This involves the use or installation (see Section 8.1.4) of the initial keying material that may be obtained during user registration. Examples include the installation of a key at a CA, trust parameters, policies, trusted parties, and algorithm preferences.

8.1.4   Keying-Material Installation Function

The security of keying-material installation is crucial to the security of a system. For this function, keying material is installed for operational use within an entity's software, hardware, system, application, cryptographic module, or device using a variety of techniques. Keying material is installed when the software, hardware, system, application, cryptographic module, or device is initially set up, when new keying material is added to the existing keying material, and when existing keying material is replaced (via re-keying, key update, or key derivation - see Section 8.2.3 and Section 8.2.4).

The process for the initial installation of keying material (e.g., by manual entry, electronic key loader, or a vendor during manufacture) **shall** include the protection of the keying material during entry into a software/hardware/system/application/device/cryptographic module, taking into account the requirements of [FIPS140] and its differing requirements for the different levels of protection, and include any additional procedures that may be required.

Many applications or systems are provided by the manufacturer with keying material that is used to test that the newly installed application/system is functioning properly. This test keying material **shall not** be used operationally.

8.1.5   Key Establishment Function

Key establishment involves the generation and distribution, or the agreement of keying material for communication between entities. All keys **shall** be generated within a FIPS140-validated cryptographic module or obtained from another source approved by the U.S. Government for the protection of national security information. During the key-establishment process, some of the keying material may be in transit (i.e., the keying material is being manually distributed or distributed using automated protocols). Other keying material may be retained locally. In either case, the keying material **shall** be protected in accordance with Section 6.

An entity may be an individual (person), organization, device or process. When keying material is generated by an entity for its own use, and the keying material is not distributed among "sub-entities" (e.g., is not distributed among various individuals, devices or processes within an organization), one or more of the appropriate protection mechanisms for stored information in Section 6.2.2 **shall** be used.

Keying material that is distributed between entities, or among an entity and its sub-entities, **shall** be protected using one or more of the appropriate protection mechanisms specified in Section 6.2.1. Any keying material that is not distributed (e.g., the private key of a key pair, or one's own copy of a symmetric key) **shall** be protected using one or more of the appropriate protection mechanisms specified in Section 6.2.2.

8.1.5.1   Generation and Distribution of Asymmetric Key Pairs

Key pairs **shall** be generated in accordance with the mathematical specifications of the appropriate **approved** Standard or NIST Recommendation.

A static key pair **shall** be generated by the entity that "owns" the key pair (i.e., the entity that uses the private key in the cryptographic computations), by a facility that distributes the key pair in accordance with Section 8.1.5.1.3, or by the user and facility in a cooperative process. When generated by the entity that owns the key pair, a signing private key **shall not** be distributed to other entities. In the case of a public signature-verification key and its associated private key, the owner **should** generate the keying material, rather than any other entity generating the keying material for that owner; this will facilitate non-repudiation.

Ephemeral keys are often used for key establishment (see [SP800-56A]). They are generated for each new key-establishment process (e.g., unique to each message or session).

The generated key pairs **shall** be protected in accordance with Section 6.1.1.

8.1.5.1.1   Distribution of Static Public Keys

Static public keys are relatively long-lived and are typically used for a number of executions of an algorithm. The distribution of the public key **should** provide assurance to the receiver of that key that the true owner of the key is known (i.e., the claimed owner is the actual owner); this requirement may be disregarded if anonymity is acceptable. However, the strength of the overall architecture and trust in the validity of the protected data depends, in large part, on the assurance of the public-key owner's identity.

In addition, the distribution of the public key **shall** provide assurance to the receiver that:

1. The purpose/usage of the key is known (e.g., RSA digital signatures or elliptic-curve key agreement),

2. Any parameters associated with the public key are known (e.g., domain parameters),

3. The public key is valid (e.g., the public key satisfies the required arithmetical properties), and

4. The owner actually possesses the corresponding private key.

8.1.5.1.1.1 Distribution of a Trust Anchor's Public Key in a PKI

The public key of a Certification Authority is the foundation for all PKI-based security services. The trust anchor is not a secret, but the *authenticity* of the trust anchor is the crucial assumption for PKI. Trust anchors may be obtained through many different mechanisms, providing different levels of assurance. The types of mechanisms that are provided may depend on the role of the user in the infrastructure. A user that is only a "relying party" – that is, a user that does not have keys registered with the infrastructure – may use different mechanisms than a user that possesses keys registered by the infrastructure. Trust anchors are frequently distributed as "self-signed" X.509 certificates, that is, certificates that are signed by the private key corresponding to the subject public key of the certificate.

Trust anchors are often embedded within an application and distributed with the application. For example, the installation of a new web browser typically includes the installation or replacement of the user's trust anchor list. Operating systems often are shipped with "code signing" trust anchor public keys. The user relies upon the authenticity of the software distribution mechanism to ensure that only valid trust anchors are installed during installation or replacement. However, in some cases other applications may install trust anchor keys in web browsers.

Trust anchors in web browsers are used for several purposes, including the validation of S/MIME e-mail certificates and web server certificates for "secure websites" that use the TLS protocol to authenticate the web server and provide confidentiality. Users who visit a "secure" website that has a certificate not issued by a trust anchor CA may be given an opportunity to accept that certificate, either for a single session or permanently. Relying users should be cautious about accepting certificates from unknown Certification Authorities so that they do not, in effect, inadvertently add new permanent trust anchors that are not trustworthy.

Warning: Roaming users **should** be aware that they are implicitly trusting all software on the host systems that they use. They should have concerns about trust anchors used by web browsers when they use systems in kiosks, libraries, Internet cafes, or hotels and systems provided by conference organizers to access "secure websites." The user has had no control over the trust anchors installed in the host system, and therefore is relying upon the host systems to have made good, sensible decisions about which trust anchors are allowed; relying parties are not participants in trust anchor selection when the trust anchors are pre-installed prior to software distribution, and may have no had no part in decisions about which trust anchors are installed thereafter. The user should be aware that he is trusting the software distribution mechanism to avoid the installation of malicious code. Extending this trust to cover trust anchors for a given application may be reasonable, and allows the relying party to obtain trust anchors without any additional procedures.

Where a user registers keys with an infrastructure, additional mechanisms are usually available. The user interacts securely with the infrastructure to register its keys (e.g., to obtain certificates), and these interactions may be extended to provide trust anchor information. This allows the user to establish trust anchor information with approximately the same assurance that the infrastructure has in the user's keys. In the case of a PKI:

1. The initial distribution of the public key of a trust anchor **should** be performed in conjunction with the presentation of a requesting entity's public key to a registration authority (RA) or CA during the certificate request process. In general, the trust anchor's public key, associated parameters, key use, and assurance of possession are conveyed as a self-signed X.509 public-key certificate. The certificate has been digitally signed by the private key that corresponds to the public key within the certificate. While the parameters and assurance of possession may be conveyed in the self-signed certificate, the trust anchor's identity and other information cannot be verified from the self-signed certificate itself (see item 2 below).

2. The trusted process used to convey a requesting entity's public key and assurances to the RA or CA **shall** also protect the trust anchor information conveyed to the requesting entity. In cases where the requesting entity appears in person, the trust anchor information may be provided at that time. If a secret value has been established during user registration (see Section 8.1.1), the trust anchor information may be supplied with the requesting entity's certificate.

8.1.5.1.1.2   Submission to a Registration Authority or Certification Authority

Public keys may be provided to a Certification Authority (CA) or to a registration authority (RA) for subsequent certification by a CA. During this process, the RA or CA **shall** obtain the assurances listed in Section 8.1.5.1.1 from the owner of the key or an authorized representative (e.g., the firewall administrator), including the owner's identity.

In general, the owner of the key is identified in terms of an identifier established during user registration (see Section 8.1.1). The key owner identifies the appropriate uses for the key, along with the parameters and any assurances of validity and possession. In cases where anonymous ownership of the public key is acceptable, the owner or the registration authority determines a pseudonym to be used as the identifier. The identifier **shall** be unique for the naming authority[28].

Proof of Possession (POP) is a mechanism that is commonly used by a CA to obtain assurance of private-key possession during key registration. In this case, the proof **shall** be provided by the reputed owner of the key pair. Without assurance of possession, it would be possible for the CA to bind the public key to the wrong entity.

The (reputed) owner **should** provide POP by performing operations with the private key that satisfy the indicated key use. For example, if a key pair is intended to support RSA key transport, the CA may provide the owner with a key that is encrypted using the owner's public key. If the owner can correctly decrypt the ciphertext key using the associated private key and then provide evidence that the key was correctly decrypted (e.g., by encrypting a random challenge from the

---

[28] The naming authority is the entity responsible for the allocation and distribution of domain names, ensuring that the names are unique within the domain. A naming authority is often restricted to a particular level of domains, such as .com, ,net or .edu.

CA), then the owner has established POP. However, when a key pair is intended to support key establishment, POP may also be afforded by using the private key to digitally sign the certificate request (although this is not the preferred method). The private key-establishment key (i.e., the private key-agreement or key-transport key) **shall not** be used to perform signature operations after certificate issuance.

As with user registration, the strength of the security infrastructure depends upon the methods used for distributing the key to an RA or CA. There are many different methods, each appropriate for some range of applications. Some examples of common methods are:

1. The public key and the information identified in Section 8.1.5.1.1 are provided by the public-key owner in person, or by an authorized representative of the public-key owner in person.

2. The identity of the public-key owner or an authorized representative of the public-key owner (i.e., a person, organization, device or process) is established at the RA or CA in person during user registration. Unique, unpredictable information (e.g., an authenticator or cryptographic key) is provided at this time by the RA or CA to the owner or authorized representative as a secret value. The public key and the information identified in Section 8.1.5.1.1 are provided to the RA or CA using a communication protocol protected by the secret value. The secret value **should** be destroyed by the key owner as specified in Section 8.3.4 upon receiving confirmation that the certificate has been successfully generated. The RA or CA may maintain this secret value for auditing purposes, but the RA or CA **should not** accept further use of the secret value to prove identity.

    When a specific list of public-key owners are pre-authorized to register keys, identifiers may be assigned without the owners being present. In this case, it is critical to protect the secret value from disclosure, and the procedures **shall** demonstrate that the chain of custody was maintained. The secret value's lifetime **should** be limited, but **shall** allow for the public-key owner to appear at the RA or CA, generate his keys, and provide the public key (under the secret value's protection) to the RA or CA. Since it may take some time for the public-key owner to appear at the RA or CA, a two or three week lifetime for the secret value is probably reasonable.

    When public-key owners are not pre-authorized, the RA or CA **shall** determine the identifier in the user's presence. In this case, the time limit may be much more restrictive, since the public-key owner need only generate his keys and provide the public key to the CA or RA. In this case, a 24-hour lifetime for the secret value would be reasonable.

3. The identity of the public-key owner is established at the RA or CA using a previous determination of the public-key owner's identity. This is accomplished by "chaining" a new public-key certificate request to a previously certified digital-signature key pair. For example, the request for a new public-key certificate is signed by the owner of the new public key to be certified. The private signature key used to sign the request **should** be associated with a public signature-verification key that is certified by the same CA that will certify the new public key. The request contains the new public key and any key-related information (e.g., key use and parameters). In addition, the CA **shall** obtain assurance of public-key validity and assurance that the owner possesses the associated private key.

4. The public key, key use, parameters, validity assurance information, and assurance of possession are provided to the RA or CA, along with a claimed identity. The RA or CA delegates the verification of the public-key owner's identity to another trusted process (e.g., an examination of the public-key owner's identity by the U.S. Postal Service when delivering registered mail containing the requested certificate). Upon receiving a request for certification, the RA or CA generates and sends unique, unpredictable information (e.g., an authenticator or cryptographic key) to the requestor using a trusted process (e.g., registered mail sent via the U.S. Postal Service). The trusted process assures that the identity of the requestor is verified prior to delivery of the information provided by the RA or CA. The owner uses this information to prove that the trusted process succeeded, and the RA or CA subsequently delivers the certificate to the owner. The information **should** be destroyed by the key owner as specified in Section 8.3.4 upon receiving confirmation that the certificate has been successfully generated. (The RA or CA may maintain this information for auditing purposes, but **should not** accept further use of the unique identifier to prove identity.)

In cases involving an RA, upon receipt of all information from the requesting entity (i.e., the owner of the new public key), the RA forwards the relevant information to a CA for certification. The RA and CA, in combination, **shall** perform any validation or other checks required for the algorithm with which the public key will be used (e.g., public-key validation) prior to issuing a certificate. The CA **should** indicate the checks or validations that have been performed (e.g., in the certificate, or in the certificate policy or certification practice statement). After generation, the certificate is distributed manually or using automated protocols to the RA, the public-key owner, or a certificate repository (i.e., a directory) in accordance with the CA's certification practice statement.

8.1.5.1.1.3 General Distribution

Public keys may be distributed to entities other than an RA or CA in several ways. Distribution methods include:

1. Manual distribution of the public key itself by the owner of the public key (e.g., in a face-to-face transfer or by a bonded courier); the mandatory assurances listed in Section 8.1.5.1.1 **shall** be provided to the recipient prior to the use of the public key operationally.

2. Manual (e.g., in a face-to-face transfer or by receipted mail) or automated distribution of a public-key certificate by the public-key owner, the CA, or a certificate repository (i.e., a directory). The mandatory assurances listed in Section 8.1.5.1.1 that are not provided by the CA (e.g., public-key validation) **shall** be provided to or performed by the receiver of the public key prior to the use of the key operationally.

3. Automated distribution of a public key (e.g., using a communication protocol with authentication and content integrity) in which a certified key pair owned by the entity distributing the public key protects the public key being distributed. The mandatory assurances listed in Section 8.1.5.1.1 **shall** be provided to the receiving entity prior to the use of the public key operationally.

8.1.5.1.2 Distribution of Ephemeral Public Keys

When used, ephemeral public keys are distributed as part of a secure key-agreement protocol. The key-agreement process (i.e., the key-agreement scheme + the protocol + key confirmation + any associated negotiation + local processing) **should** provide a recipient with the assurances listed in Section 8.1.5.1.1. The recipient of an ephemeral public key **shall** obtain assurance of validity of that key as specified in [SP800-56A] or [SP800-56B] prior to using that key for subsequent steps in the key-agreement process.

8.1.5.1.3 Distribution of Centrally Generated Key Pairs

When a static key pair is centrally generated, the key pair **shall** be generated within a FIPS140-validated cryptographic module or obtained from another source approved by the U.S. government for protecting national security information for subsequent delivery to the intended owner of the key pair. A signing key pair generated by a central key-generation facility for its subscribers will not provide strong non-repudiation for those individual subscribers; therefore, when non-repudiation is required by those subscribers, the subscribers **should** generate their own signing key pairs. However, if the central key-generation facility generates signing key pairs for its own organization and distributes them to members of the organization, then non-repudiation may be provided at an organizational level (but not an individual level).

The private key of a key pair generated at a central facility **shall** only be distributed to the intended owner of the key pair. The confidentiality of the centrally generated private key **shall** be protected, and the procedures for distribution **shall** include an authentication of the recipient's identity as established during user registration (see Section 8.1.1).

The key pair may be distributed to the intended owner using an appropriate manual method (e.g., courier, mail or other method specified by the key-generation facility) or secure automated method (e.g., a secure communication protocol). The private key **shall** be distributed in the same manner as a symmetric key (see Section 8.1.5.2.2). During the distribution process, each key of the key pair **shall** be provided with the appropriate protections for that key (see Section 6.1).

When split-knowledge procedures are used for the manual distribution of the private key, the key **shall** be split into multiple key components that have the same security properties as the original key (e.g., randomness); each key component **shall** provide no knowledge of the value of the original key (e.g., each key component **shall** appear to be generated randomly).

Upon receipt of the key pair, the owner **shall** obtain assurance of the validity of the public key (see [SP800-56A], [SP800-56B] and [SP800-89]. The owner **shall** obtain assurance that the public and private keys of the key pair are correctly associated (i.e., check that they are a consistent pair, for example, by checking that a key encrypted under a pubic key-transport key can be decrypted by the private key-transport key).

8.1.5.2 Generation and Distribution of Symmetric Keys

The symmetric keys used for the encryption and decryption of data or other keys and for the computation of MACs (see Sections 4.2.2 and 4.2.3) **shall** be determined by an **approved** method and **shall** be provided with protection that is consistent with Section 6.

Symmetric keys **shall** be either:

1. Generated and subsequently distributed (see Sections 8.1.5.2.1 and 8.1.5.2.2) either manually (see Section 8.1.5.2.2.1), using a public key-transport mechanism (see Section

8.1.5.2.2.2), or using a previously distributed or agreed-upon key wrapping key (see Section 8.1.5.2.2.2),

2. Established using a key-agreement scheme (i.e., the generation and distribution are accomplished with one process) (see Section 8.1.5.2.3),

3. Determined by a key-update process (see Section 8.2.3.2), or

4. Derived from a master key (see Section 8.2.4).

8.1.5.2.1 Key Generation

Symmetric keys determined by key generation methods **shall** be either generated by an **approved** method (e.g., using an **approved** random number generator), created from the previous key during a key update procedure (see Section 8.2.3.2), or derived from a master key (see Section 8.2.4) using an **approved** key-derivation function (see [SP800-108]). Also, see [SP800-133].

When split-knowledge procedures are used, the key **shall** exist outside of a [FIPS140] cryptographic module as multiple key components. The keying material may be created within a cryptographic module and then split into components for export from the module, or may be created as separate components. Each key component **shall** provide no knowledge of the key value (e.g., each key component must appear to be generated randomly). If knowledge of k (where k is less than or equal to n) components is required to construct the original key, then knowledge of any k-1 key components **shall** provide no information about the original key other than, possibly, its length. Note: A suitable combination function is not provided by simple concatenation; e.g., it is not acceptable to form an 80-bit key by concatenating two 40-bit key components.

All keys and key components **shall** be generated within a FIPS140-validated cryptographic module or obtained from another source approved by the U.S. Government for the protection of national security information.

8.1.5.2.2 Key Distribution

Keys generated in accordance with Section 8.1.5.2.1 as key-wrapping keys (i.e., key encrypting keys), as the initial key for key update, as master keys to be used for key derivation, or for the protection of communicated information are distributed manually (manual key transport) or using an automated key-transport protocol (automated key transport).

Keys used only for the storage of information (i.e., data or keying material) **shall not** be distributed except for backup or to other authorized entities that may require access to the information protected by the keys.

8.1.5.2.2.1 Manual Key Distribution

Keys distributed manually (i.e., by other than an automated key-transport protocol) **shall** be protected throughout the distribution process. During manual distribution, secret or private keys **shall** either be wrapped (i.e., encrypted) or be distributed using appropriate physical security procedures. If multi-party control is desired, split knowledge procedures may be used as well. The manual distribution process **shall** assure that:

1. The distribution of keys is from an authorized source,

2. Any entity distributing plaintext keys is trusted by both the entity that generates the keys and the entity(ies) that receives the keys,

3. The keys are protected in accordance with Section 6, and

4. The keys are received by the authorized recipient.

When distributed in encrypted form, the key **shall** be encrypted by an **approved** key-wrapping scheme using a key-wrapping key that is used only for key wrapping, or by an **approved** key-transport scheme using a public key-transport key owned by the intended recipient. The key-wrapping key or public key-transport key **shall** have been distributed as specified in this Recommendation.

When using split knowledge procedures, each key component **shall** be either encrypted or distributed separately to each individual. Appropriate physical security procedures **shall** be used to protect each key component as sensitive information.

Physical security procedures may be used for all forms of manual key distribution. However, these procedures are particularly critical when the keys are distributed in plaintext form. In addition to the assurances listed above, accountability and auditing of the distribution process (see Sections 9.1 and 9.2) **should** be used.

8.1.5.2.2.2   Automated Key Distribution/Key Transport

Automated key distribution, or key transport, is used to distribute keys via a communication channel (e.g., the Internet or a satellite transmission). Automated key-transport requires the prior distribution of a key-wrapping key (i.e., a key-encryption key) or a public key-transport key as follows:

1. A key-wrapping key **shall** be generated and distributed in accordance with Sections 8.1.5.2.1 and 8.1.5.2.2, or established using a key-agreement scheme as defined in Section 8.1.5.2.3.

2. A public key-transport key **shall** be generated and distributed as specified in Section 8.1.5.1.

Only **approved** key-wrapping key or public key-transport schemes **shall** be used. The **approved** key-transport schemes provide assurance that:

a. For symmetric key-wrapping schemes: The key-wrapping key and the distributed key are not disclosed or modified.

b. For public key-transport schemes: The private key-transport key and the distributed key are not disclosed or modified, and correct association between the private and public key-transport keys is maintained.

c. The keys are protected in accordance with Section 6.

In addition, the **approved** key-transport schemes, together with the associated key-establishment protocol, **should** provide the following assurances:

d. Each entity in the key-transport process knows the identifier associated with the other entity(ies),

e. The keys are correctly associated with the entities involved in the key-transport process, and

f. The keys have been received correctly.

8.1.5.2.3 Key Agreement

Key agreement is used in a communication environment to establish keying material using information contributed by all entities in the communication (most commonly, only two entities) without actually sending the keying material. Only **approved** key-agreement schemes **shall** be used. **Approved** key-agreement schemes using asymmetric techniques are provided in [SP800-56A] and [SP800-56B]. Key agreement uses asymmetric key pairs or symmetric key-encrypting keys (i.e., key-wrapping keys) to calculate shared secrets, which are then used to derive symmetric keys and other keying material (e.g., IVs).

A key-agreement scheme uses (1) symmetric key-encrypting keys, or (2) either static or ephemeral key pairs or both. The asymmetric key pairs **should** be generated and distributed as discussed in Section 8.1.5.1. Keying material derived from a key-agreement scheme **shall** be protected as specified in Section 6.

A key-agreement scheme and its associated key-establishment protocol **should** provide the following assurances:

1. Identifiers for entities involved in the key establishment protocol are correctly associated with those entities. Assurance for the association of identifiers to entities may be achieved by the key-agreement scheme or may be achieved by the protocol in which key agreement is performed. Note that the identifier may be a "pseudo-identifier", not the identifier appearing on the entity's birth certificate, for example.

   In the general case, an identifier is associated with each party involved in the key-establishment protocol, and each entity in the key-establishment process must be able to associate all the other entities with their appropriate identifier. In special cases, such as the secure distribution of public information on a web site, the association with an identifier may only be required for a subset of the entities (e.g., only the server).

2. The keys used in the key-agreement scheme are correctly associated with the entities involved in the key-establishment process.

3. The derived keys are correct.

Keys derived through key agreement and its enabling protocol **should not** be used to protect and send information until the three assurances described above have been achieved.

8.1.5.3 Generation and Distribution of Other Keying Material

Keys are often generated in conjunction with or are used with other keying material. This other keying material **shall** be protected in accordance with Section 6.2.

8.1.5.3.1 Domain Parameters

Domain parameters are used by some public-key algorithms to generate key pairs, to compute digital signatures, or to establish keys. Typically, domain parameters are generated infrequently and used by a community of users for a substantial period of time. Domain parameters may be distributed in the same manner as the public keys with which they are associated, or they may be made available at some other accessible site. Assurance of the validity of the domain parameters **shall** be obtained prior to use, either by a trusted entity that vouches for the parameters (e.g., a CA), or by the entities themselves. Assurance of domain-parameter validity is addressed in

[SP800-89] and [SP800-56A]. Obtaining this assurance **should** be addressed in a CA's certification practices statement or an organization's security plan.

8.1.5.3.2   Initialization Vectors

Initialization vectors (IVs) are used by symmetric algorithms in several modes of operation for encryption and decryption, or for authentication. The criteria for the generation and use of IVs are provided in [SP800-38A]; IVs **shall** be protected as specified in Section 6. When distributed, IVs may be distributed in the same manner as their associated keys, or may be distributed with the information that uses the IVs as part of the encryption or authentication mechanism.

8.1.5.3.3   Shared Secrets

Shared secrets are computed during a key-agreement process and are subsequently used to derive keying material. Shared secrets are generated as specified by the appropriate key-agreement scheme (see [SP800-56A] and [SP800-56B]), but **shall not** be used as keying material.

8.1.5.3.4   RNG Seeds

Seeds are used to initialize a Deterministic Random Bit Generator (DRBG). The criteria for the selection of a seed for an RNG are provided in the specification of an **approved** DRBG (e.g., see [SP800-90A]). The seeds for an RNG consist of a string of bits containing entropy (i.e., entropy input) and may possibly include "other information" that may be either public or secret (see [SP800-90A]). The entropy input **shall** be destroyed immediately after use; however, all or a portion of the "other information" may be reused. Any portion of the "other information" that is public **shall** be protected as "other public information" (see Table 6); any portion of the "other information" that is secret **shall** be handled as "other secret information" (see Table 6). In this document, the term "RNG seed" will be used as a collective term for the entropy input and any other secret information that is used in the DRBG seeding process.

When entropy input or other secret information is distributed for seeding a DRBG, it **shall** be distributed using a secure channel that protects the integrity and confidentiality of the distributed material. When entropy input is distributed, the entity that distributes the entropy input **shall** destroy its copy of the entropy input immediately after distributing it.

8.1.5.3.5   Other Public and Secret Information

Public and secret information may be used during the seeding of an RNG (see Section 8.1.5.3.4) or during the generation or establishment of keying material (see [SP800-56A], [SP800-56B] and [SP800-108]). Public information may be distributed; secret information **shall** be protected in the same manner as a private or secret key during distribution.

8.1.5.3.6   Intermediate Results

Intermediate results occur during computation using cryptographic algorithms. These results **shall not** be distributed as or with the keying material.

8.1.5.3.7   Random Numbers

Random numbers are used for many purposes, including the generation of keys and nonces, and the issuing of challenges during communication protocols. Random numbers may be distributed, but whether or not confidentiality protection is required depends on the context in which the random number is used.

8.1.5.3.8 Passwords

Passwords are used to verify authentication or authorization, and, in some cases, to derive keying material (see [SP800-132]). Passwords may be distributed, but their protection during distribution **shall** be consistent with the protection required for their use. For example, if the password will be used to access cryptographic keys that are used to provide 128 bits of security strength when protecting data, then the password needs to be provided with at least 128 bits of protection as well. Note that poorly selected passwords may not themselves provide the required amount of protection for key access and are potentially the weak point of the process; i.e., it may be far easier to guess the password than to attempt to "break" the cryptographic protection used on the password. It is the responsibility of users and organizations to select passwords that provide the requisite amount of protection for the keys they protect.

8.1.6   Key Registration Function

Key registration results in the binding of keying material to information or attributes associated with a particular entity. Keys that would be registered include the public key of an asymmetric key pair and the symmetric key used to bootstrap an entity into a system. Normally, keys generated during communications (e.g., using key-agreement schemes or key derivation functions) would not be registered. Information provided during registration typically includes the identifier of the entity associated with the keying material and the intended use of the keying material (e.g., as a signing key, data-encryption key, etc.). Additional information may include authorization information or specify the level of trust. The binding is performed after the entity's identity has been authenticated by a means that is consistent with the system policy (see Section 8.1.1). The binding provides assurance to the community-at-large that the keying material is used by the correct entity in the correct application. The binding is often cryptographic, which creates a strong association between the keying material and the entity. A trusted third party performs the binding. Examples of a trusted third party include a Kerberos realm server or a PKI certification authority (CA). Identifiers issued by a trusted third party **shall** be unique to that party.

When a Kerberos realm server performs the binding, a symmetric key is stored on the server with the corresponding attributes. In this case, the registered keying material is maintained in confidential storage (i.e., the keys are provided with confidentiality protection).

When a CA performs the binding, the public key and associated attributes are placed in a public-key certificate, which is digitally signed by the CA. In this case, the registered keying material may be publicly available.

When a CA provides a certificate for a public key, the public key **shall** be verified to ensure that it is associated with the private key known by the purported owner of the public key. This provides assurance of possession. When POP is used to obtain assurance of possession, the assurance **shall** be accomplished as specified in Section 8.1.5.1.1.2.

8.2   Operational Phase

Keying material used during the cryptoperiod of a key is often stored for access as needed. During storage, the keying material **shall** be protected as specified in Section 6.2.2. During normal use, the keying material is stored either on the device or module that uses that material, or on an immediately accessible storage media. When the keying material is required for operational use, the keying material is acquired from immediately accessible storage when not present in active memory within the device or module.

To provide continuity of operations when the keying material becomes unavailable for use from normal operational storage during its cryptoperiod (e.g., because the material is lost or corrupted), keying material may need to be recoverable. If an analysis of system operations indicates that the keying material needs to be recoverable, then the keying material **shall** either be backed up (see Section 8.2.2.1), or the system **shall** be designed to allow reconstruction (e.g., re-derivation) of the keying material. Retrieving or reconstructing keying material from backup or an archive is commonly known as key recovery (see Section 8.2.2.2).

At the end of a key's cryptoperiod, a new key needs to be available to replace the old key if operations are to be continued. This can be accomplished by re-keying (see Section 8.2.3.1), key update (see Section 8.2.3.2), or key derivation (see Section 8.2.4). A key **shall** be destroyed in accordance with Section 8.3.4 and **should** be destroyed as soon as that key is no longer needed in order to reduce the risk of exposure.

8.2.1 Normal Operational Storage Function

The objective of key management is to facilitate the operational availability of keying material for standard cryptographic purposes. Usually, a key remains operational until the end of the key's cryptoperiod (i.e., the expiration date). During normal operational use, keying material is available either in the device or module (e.g., in RAM) or in an immediately accessible storage media (e.g., on a local hard disk).

8.2.1.1 Device or Module Storage

Keying material may be stored in the device or module that adds, checks, or removes the cryptographic protection on information. The storage of the keying material **shall** be consistent with Section 6.2.2, as well as with [FIPS140].

8.2.1.2 Immediately Accessible Storage Media

Keying material may need to be stored for normal cryptographic operations on an immediately accessible storage media (e.g., a local hard drive) during the cryptoperiod of the key. The storage requirements of Section 6.2.2 **shall** apply to this keying material.

8.2.2 Continuity of Operations Function

Keying material can become lost or unusable, due to hardware damage, corruption or loss of program or data files, or system policy or configuration changes. In order to maintain the continuity of operations, it is often necessary for users and/or administrators to be able to recover keying materials from backup storage. However, if operations can be continued without the backup of keying material (e.g., by re-keying), or the keying material can be recovered or reconstructed without being saved, it may be preferable not to save the keying material in order to lessen the possibility of a compromise of the keying material or other cryptographically related information.

The compromise of keying material affects continuity of operations (see Section 8.4). When keying material is compromised, continuity of operations requires the establishment of entirely new keying material (see Section 8.1.5), following an assessment of what keying material is affected and needs to be replaced.

8.2.2.1   Backup Storage

The backup of keying material on an independent, secure storage media provides a source for key recovery (see Section 8.2.2.2). Backup storage is used to store copies of information that is also currently available in normal operational storage during a key's cryptoperiod (i.e., in the cryptographic device or module, or on an immediately accessible storage media - see Section 8.2.1.1). Not all keys need be backed up. The storage requirements of Section 6.2.2 apply to keying material that is backed up. Tables 7 and 8 provide guidance about the backup of each type of keying material and other related information. An "OK" indicates that storage is permissible, but not necessarily required. The final determination for backup **should** be made based on the application in which the keying material is used. A detailed discussion about each type of key and other cryptographic information is provided in Appendix B.3.

Keying material maintained in backup **should** remain in storage for at least as long as the same keying material is maintained in storage for normal operational use (see Section 8.2.1). When no longer needed for normal operational use, the keying material and other related information **should** be removed from backup storage. When removed from backup storage, all traces of the information in backup storage **shall** be destroyed in accordance with Section 8.3.4.

A discussion of backup and recovery is provided in [ITLBulletin].

Table 7: Backup of keys

Type of Key	Backup?
Private signature key	No (in general); non-repudiation would be in question. However, it may be warranted in some cases - a CA's private signing key, for example. When required, any backed up keys **shall** be stored under the owner's control.
Public signature-verification key	OK; its presence in a public-key certificate that is available elsewhere may be sufficient.
Symmetric authentication key	OK
Private authentication key	OK, if required by an application.
Public authentication key	OK; if required by an application.
Symmetric data encryption key	OK
Symmetric key-wrapping key	OK
Random number generation key	Not necessary and may not be desirable, depending on the application.
Symmetric master key	OK
Private key-transport key	OK
Public key-transport key	OK; its presence in a public-key certificate that is available elsewhere may be sufficient.
Symmetric key-agreement key	OK
Private static key-agreement key	OK
Public static key-agreement key	OK; its presence in a public-key certificate that is available elsewhere may be sufficient.
Private ephemeral key-agreement key	No
Public ephemeral key-agreement key	OK
Symmetric authorization key	OK
Private authorization key	OK
Public authorization key	OK; its presence in a public-key certificate that is available elsewhere may be sufficient.

Table 8: Backup of other cryptographic or related information

Type of Keying Material	Backup?
Domain parameters	OK
Initialization vector	OK, if necessary
Shared secret	No
RNG seed	No
Other public information	OK
Other secret information	OK
Intermediate results	No
Key control information (e.g., IDs, purpose, etc.)	OK
Random number	Depends on application or use of the random number
Passwords	OK when used to derive keys or to detect the reuse of passwords; otherwise, No
Audit information	OK

8.2.2.2   Key Recovery Function

Keying material that is in active memory or stored in normal operational storage may sometimes be lost or corrupted (e.g., from a system crash or power fluctuation). Some of the keying material is needed to continue operations and cannot easily be replaced. An assessment needs to be made of which keying material needs to be preserved for possible recovery at a later time.

The decision as to whether key recovery is required **should** be made on a case-by-case basis. The decision **should** be based on:

1. The type of key (e.g., private signature key, symmetric data-encryption key),

2. The application in which the key will be used (e.g., interactive communications, file storage),

3. Whether the key is "owned" by the local entity (e.g., a private key) or by another entity (e.g., the other entity's public key) or is shared (e.g., a symmetric data-encryption key shared by two entities),

4. The role of the entity in a communication (e.g., sender or receiver), and

5. The algorithm or computation in which the key will be used (e.g., does the entity have the necessary information to perform a given computation if the key were to be recovered)[29].

---

[29] This could be the case when performing a key-establishment process for some key-establishment schemes (see SP 800-56A and SP 800-56B).

The factors involved in a decision for or against key recovery **should** be carefully assessed. The trade-offs are concerned with a continuity of operations versus the risk of possibly exposing the keying material and the information it protects if control of the keying material is lost. If it is determined that a key needs to be recovered, and the key is still active (i.e., the cryptoperiod of the key has not expired), then the key may be replaced in order to limit the exposure of the data protected by that key (see Section 8.2.3).

Issues associated with key recovery and discussions about whether or not different types of cryptographic material need to be recoverable are provided in Appendix B.

8.2.3   Key Change Function

Key change is the replacement of a key with another key that performs the same function as the original key. There are several reasons for changing a key.

1. The key may have been compromised.

2. The key's cryptoperiod may be nearing expiration.

3. It may be desirable to limit the amount of data protected with any given key.

A key may be replaced by re-keying or by key update.

8.2.3.1   Re-keying

If the new key is generated in a manner that is entirely independent of the "value" of the old key, the process is known as re-keying. This replacement **shall** be accomplished using one of the key-establishment methods discussed in Section 8.1.5. Re-keying is used when a key has been compromised (provided that the re-keying scheme itself is not compromised) or when the cryptoperiod is nearing expiration.

8.2.3.2   Key Update Function

If the "value" of the new key is dependent on the value of the old key, the process is known as key update (i.e., the current key is modified to create a new key). This **shall** be accomplished by applying a non-reversible function to the old key and possibly other data. Unlike re-keying, key update may not require the exchange of any new information between the entities that previously shared the old key. For example, the two entities may agree to update their shared key on the first day of each month. Since a non-reversible function is used in the update process, previous keys are protected in the event that a key is compromised. However, future keys are not protected. After a limited number of updates, new keying material **shall** be established by employing a fresh re-key operation (see Section 8.2.3.1). Key update is often used to limit the amount of data protected by a single key, but it **shall not** be used to replace a compromised key.

8.2.4   Key Derivation Function

Cryptographic keys may be derived from a secret value. The secret value, together with other information, is input into a key-derivation method (e.g., a key-derivation function) that outputs the required key(s). In contrast to key change, the derived keys are often used for new purposes, rather than for replacing the secret values from which they are derived. The derivation method **shall** be non-reversible so that the secret value cannot be determined from the derived keys. In addition, it **shall not** be possible to determine a derived key from other derived keys. It should be noted that the strength of a derived key is no greater than the strength of the derivation algorithm and the secret value from which the key is derived.

Three commonly used key-derivation cases are discussed below.

1. *Two parties derive common keys from a common shared secret.* This approach is used in the key-establishment techniques specified in [SP800-56A] and [SP800-56B]. The security of this process is dependent on the security of the shared secret and the specific key-derivation method used. If the shared secret is known, the derived keys may be determined. A key-derivation method specified in [SP800-56A], [SP800-56B] or [SP800-56C] **shall** be used for this purpose. These derived keys may be used to provide the same confidentiality, authentication, and data integrity services as randomly generated keys, with a security strength determined by the scheme and key pairs used to generate the shared secret.

2. *Keys derived from a key-derivation key (master key).* This is often accomplished by using the key-derivation key, entity ID, and other known information as input to a function that generates the keys. One of the key-derivation functions defined in [SP800-108] **shall** be used for this purpose. The security of this process depends upon the security of the key-derivation key and the key-derivation function. If the key-derivation key is known by an adversary, he can generate any of the derived keys. Therefore, keys derived from a key-derivation key are only as secure as the key-derivation key itself. As long as the key-derivation key is kept secret, the derived keys may be used in the same manner as randomly generated keys.

3. *Keys derived from a password.* A user-generated password, by its very nature, is less random (i.e., has lower entropy) than is required for a cryptographic key; that is, the number of passwords that are likely to be used to derive a key is significantly smaller than the number of keys that are possible for a given key size. In order to increase the difficulty of exhaustively searching the likely passwords, a key-derivation function is iterated a large number of times. The key is derived using a password, entity ID, and other known information as input to the key-derivation function. The security of the derived key depends upon the security of the password and the key-derivation process. If the password is known or can be guessed, then the corresponding derived key can be generated. Therefore, keys derived in this manner are likely to be less secure than randomly generated keys, or keys derived from a shared secret or key-derivation key. For storage applications, one of the key-derivation methods specified in [SP800-132] **shall** be used to derive keys. For non-storage applications, keys derived in this manner **shall** be used for authentication purposes only and not for general encryption.

8.3 Post-Operational Phase

During the post-operational phase, keying material is no longer in operational use, but access to the keying material may still be possible.

8.3.1 Archive Storage and Key Recovery Functions

An archive for keying material **shall** provide both integrity and access control. Integrity is required in order to protect the archived material from unauthorized modification, deletion, and insertion. Access control is needed to prevent unauthorized disclosure. Archived information **shall** be protected as specified in Section 6.2.2. When keying material is entered into the archive, it is often timestamped so that the date-of-entry can be determined. This date may itself be cryptographically protected so that it cannot be changed without detection.

If keying material needs to be recoverable (e.g., after the end of its cryptoperiod), either the keying material **shall** be archived, or the system **shall** be designed to allow reconstruction (e.g., re-derivation) of the keying material from archived information. Retrieving the keying material from archive storage or by reconstruction is commonly known as key recovery. The archive **shall** be maintained by a trusted party (e.g., the organization associated with the keying material or a trusted third party).

A key management archive is a repository containing keying material and other related information for recovery as needed. Not all keying material needs to be archived. An organization's security plan **should** indicate the types of information that are to be archived (see Part 2).

While in storage, archived information may be either static (i.e., never changing) or may need to be re-encrypted under a new archive-encryption key from time-to-time. Archived data **should** be stored separately from operational data, and multiple copies of archived cryptographic information **should** be provided in physically separate locations (i.e., it is recommended that the key management archive be backed up). For critical information that is encrypted under archived keys, it may be necessary to back up the archive keys and to store multiple copies of these archive keys in separate locations.

When archived, keying material **should** be archived prior to the end of the cryptoperiod of the key. For example, it may be prudent to archive the keying material during key activation. When no longer required, the keying material **shall** be destroyed in accordance with Section 8.3.4.

Archived cryptographic information requires protection in accordance with Section 6.2.2. Confidentiality is provided by an archive-encryption key (one or more encryption keys that are used exclusively for the encryption of archived information), by another key that has been archived, or by a key that may be derived from an archived key. When encrypted by the archive-encryption key, the encrypted keying material **shall** be re-encrypted by any new archive-encryption key at the end of the cryptoperiod of the old archive-encryption key. When the keying material is re-encrypted, integrity values on that keying material **shall** be recomputed. This imposes a significant burden; therefore, the strength of the cryptographic algorithm **shall** be selected to minimize the need for re-encryption.

Likewise, integrity protection may be provided by an archive-integrity key (one or more authentication or digital-signature keys that are used exclusively for the archive) or by another key that has been archived. If integrity protection is to be maintained at the end of the cryptoperiod of the archive-integrity key, new integrity values **shall** be computed on the archived information on which the old archive-integrity key was applied.

The archive-encryption and archive-integrity keys may be either symmetric keys or public-key pairs. Unless the cryptographic algorithm is specifically designed to provide both integrity and confidentiality with a single key, the keys used for confidentiality and integrity **shall** be different, and **shall** be protected in the same manner as their key type (see Section 6).

Tables 9 and 10 indicate the appropriateness of archiving keys and other cryptographically related information. An "OK" in column 2 (Archive?) indicates that archival is permissible, but not necessarily required. Column 3 (Retention period) indicates the minimum time that the key

should be retained in the archive. Additional advice on the storage of keying material in archive storage is provided in Appendix B.3.

Table 9: Archive of keys

Type of Key	Archive?	Retention period (minimum)
Private signature key	No	
Public signature-verification key	OK	Until no longer required to verify data signed with the associated private key
Symmetric authentication key	OK	Until no longer needed to authenticate data or an identity.
Private authentication key	No	
Public authentication key	OK	
Symmetric data-encryption key	OK	Until no longer needed to decrypt data encrypted by this key
Symmetric key-wrapping key	OK	Until no longer needed to decrypt keys encrypted by this key
Symmetric random number generator key	No	
Symmetric master key	OK, if needed to derive other keys for archived data	Until no longer needed to derive other keys
Private key-transport key	OK	Until no longer needed to decrypt keys encrypted by this key
Public key-transport key	OK	
Symmetric key-agreement key	OK	
Private static key-agreement key	OK	
Public static key-agreement key	OK	Until no longer needed to reconstruct keying material.
Private ephemeral key-agreement key	No	
Public ephemeral key-agreement key	OK	
Symmetric authorization key	No	
Private authorization key	No	
Public authorization key	OK	

Table 10: Archive of other cryptographic related information

Type of Key	Archive?	Retention period (minimum)
Domain parameters	OK	Until all keying material, signatures and signed data using the domain parameters are removed from the archive
Initialization vector	OK; normally stored with the protected information	Until no longer needed to process the protected data
Shared secret	No	
RNG seed	No	
Other public information	OK	Until no longer needed to process data using the public information
Other secret information	OK	Until no longer needed to process data using the secret information
Intermediate result	No	
Key control information (e.g., IDs, purpose)	OK	Until the associated key is removed from the archive
Random number		Depends on application or use of the random number
Password	OK when used to derive keys or to detect the reuse of passwords; otherwise, No	Until no longer needed to (re-)derive keys or to detect password reuse
Audit information	OK	Until no longer needed

The recovery of archived keying material may be required to remove (e.g., decrypt) or check (e.g., verify a digital signature or a MAC) the cryptographic protections on other archived data. The key recovery process results in retrieving or reconstructing the desired keying material from archive storage in order to perform the required cryptographic operation. Immediately after completing this operation, the keying material **shall** be erased from the cryptographic process, but **shall** be retained in the archive (see Section 8.3.4) as long as needed. Further advice on key recovery issues is provided in Appendix B.

8.3.2   Entity De-registration Function

The entity de-registration function removes the authorizations of an entity to participate in a security domain. When an entity ceases to be a member of a security domain, the entity **shall** be de-registered. De-registration is intended to prevent other entities from relying on or using the de-registered entity's keying material.

All records of the entity and the entity's associations **shall** be marked to indicate that the entity is no longer a member of the security domain, but the records **should not** be deleted. To reduce confusion and unavoidable human errors, identification information associated with the de-registered entity **should not** be re-used (at least for a period of time). For example, if a "John Wilson" retires and is de-registered on Friday, the identification information assigned to his son "John Wilson", who is hired the following Monday, **should** be different.

8.3.3   Key De-registration Function

Registered keying material may be associated with the identity of a key owner, owner attributes (e.g., email address), role or authorization information. When the keying material is no longer needed, or the associated information becomes invalid, the keying material **should** be de-registered (i.e., all records of the keying material and its associations **should** be marked to indicate that the key is no longer in use) by the appropriate trusted third party. In general, this will be the trusted third party that registered the key (see Section 8.1.6).

Keying material **should** be de-registered when the attributes associated with an entity are modified. For example, if an entity's email address is associated with a public key, and the entity's address changes, the keying material **should** be de-registered to indicate that the associated attributes have become invalid. Unlike the case of key compromise, the entity could safely re-register the public key after modifying the entity's attributes through the user registration process (see Section 8.1.1).

When a registered cryptographic key is compromised, that key and any associated keying material **shall** be de-registered. When the compromised key is the private part of a public-private key pair, the public key **shall** also be revoked (see Section 8.3.5). If the attributes associated with a public-private key pair are changed, but the private key has not been compromised, the public key **should** be revoked with an appropriate reason code (see Section 8.3.5).

8.3.4   Key Destruction Function

When copies of cryptographic keys are made, care should be taken to provide for their eventual destruction. All copies of the private or symmetric key **shall** be destroyed as soon as they are no longer required (e.g., for archival or reconstruction activity) in order to minimize the risk of a compromise. Any media on which unencrypted keying material requiring confidentiality protection is stored **shall** be destroyed in a manner that removes all traces of the keying material so that it cannot be recovered by either physical or electronic means[30]. Public keys may be retained or destroyed, as desired.

---

[30] A simple deletion of the keying material might not completely obliterate the information. For example, erasing the information might require overwriting that information multiple times with other non-related information, such as random bits, or all zero or one bits. Keys stored in memory for a long time can become "burned in". This can be mitigated by splitting the key into components that are frequently updated (see [DiCrescenzo]).

8.3.5   Key Revocation Function

It is sometimes necessary to remove keying material from use prior to the end of its normal cryptoperiod for reasons that include key compromise, removal of an entity from an organization, etc. This process is known as key revocation and is used to explicitly revoke a symmetric key or the public key of a key pair, although the private key associated with the public key is also revoked.

Key revocation may be accomplished using a notification indicating that the continued use of the keying material is no longer recommended. The notification could be provided by actively sending the notification to all entities that might be using the revoked keying material, or by allowing the entities to request the status of the keying material (i.e., a "push" or a "pull" of the status information). The notification **should** include a complete identification of the keying material, the date and time of revocation and the reason for revocation, when appropriate (e.g., key compromised). Based on the revocation information provided, other entities could then make a determination of how they would treat information protected by the revoked keying material.

For example, if a public signature-verification key is revoked because an entity left an organization, it may be appropriate to honor all signatures created prior to the revocation date. If a signing private key is compromised, resulting in the revocation of the associated public key, an assessment needs to be made as to whether or not information signed prior to the revocation would be considered as valid.

As another example, a symmetric key that is used to generate MACs may be revoked so that it will not be used to generate MACs on new information. However, the key may be retained so that archived documents can be verified.

The details for key revocation **should** reflect the lifecycle for each particular key. If a key is used in a pair-wise situation (e.g., two entities communicating in a secure session), the entity revoking the key **shall** inform the other entity of the revocation. If the key has been registered with an infrastructure, the entity revoking the key cannot always directly inform the other entities that may rely upon that key. Instead, the entity revoking the key **shall** inform the infrastructure that the key needs to be revoked (e.g., using a certificate revocation request). The infrastructure **shall** respond by de-registering the key material (see 8.3.3).

In a PKI, key revocation is commonly achieved by including the certificate in a list of revoked certificates (i.e., a CRL). If the PKI uses online status mechanisms (e.g., the Online Certificate Status Protocol [RFC 2560]), revocation is achieved by informing the appropriate certificate status server(s). For example, when a private key is compromised, the corresponding public-key certificate **should** be revoked. Certificate revocation because of a key compromise indicates that the binding between the owner and the key is no longer to be trusted. Other revocation reasons indicate that the original binding is invalid, but the private key was not compromised.

In a symmetric-key system, key revocation could, in theory, be achieved by simply deleting the key from the server's storage. Key revocation for symmetric keys is more commonly achieved by adding the key to a blacklist or compromised key list; this helps satisfy auditing and management requirements.

8.4   Destroyed Phase

The keying material is no longer available. All records of its existence may have been deleted. However, some organizations may require the retention of certain key attributes for audit purposes. For example, if a copy of an ostensibly destroyed key is found in an uncontrolled environment or is later determined to have been compromised, records of the identifier of the key, its type, and its cryptoperiod may be helpful in determining what information was protected under the key and how best to recover from the compromise.

In addition, by keeping a record of the attributes of both destroyed and destroyed compromised keys, one will be able to track which keys transitioned through a normal lifecycle and which ones were compromised at some time during their lifecycle. Thus, protected information that is linked to key names that went through the normal lifecycle may still be considered secure, provided that the security strength of the algorithm remains sufficient. However, any protected information that is linked to a key name that has been compromised may itself be compromised.

9   Accountability, Audit, and Survivability

Systems that process valuable information require controls in order to protect the information from unauthorized disclosure and modification. Cryptographic systems that contain keys and other cryptographic information are especially critical. Three useful control principles and their application to the protection of keying material are highlighted in this section.

9.1   Accountability

Accountability involves the identification of those entities that have access to, or control of, cryptographic keys throughout their lifecycles. Accountability can be an effective tool to help prevent key compromises and to reduce the impact of compromises once they are detected. Although it is preferred that no humans are able to view keys, as a minimum, the key management system **should** account for all individuals who are able to view plaintext cryptographic keys. In addition, more sophisticated key-management systems may account for all individuals authorized to access or control any cryptographic keys, whether in plaintext or ciphertext form. For example, a sophisticated accountability system might be able to determine each individual who had control of any given key over its entire lifespan. This would include the person in charge of generating the key, the person who used the key to cryptographically protect data, and the person who was responsible for destroying the key when it was no longer needed. Even though these individuals never actually saw the key in plaintext form, they are held accountable for the actions that they performed on or with the key.

Accountability provides three significant advantages:

1. It aids in the determination of when the compromise could have occurred and what individuals could have been involved,

2. It tends to protect against compromise, because individuals with access to the key know that their access to the key is known, and

3. It is very useful in recovering from a detected key compromise to know where the key was used and what data or other keys were protected by the compromised key.

Certain principles have been found to be useful in enforcing the accountability of cryptographic keys. These principles might not apply to all systems or all types of keys. Some of the principles apply to long-term keys that are controlled by humans. The principles include:

a. Uniquely identifying keys,

b. Identifying the key user,

c. Identifying the dates and times of key use, along with the data that is protected, and

d. Identifying other keys that are protected by a symmetric or private key.

9.2 Audit

Two types of audit **should** be performed on key management systems:

1. The security plan and the procedures that are developed to support the plan **should** be periodically audited to ensure that they continue to support the Key Management Policy (see Part 2).

2. The protective mechanisms employed **should** be periodically reassessed with respect to the level of security that they provide and are expected to provide in the future, and that the mechanisms correctly and effectively support the appropriate policies. New technology developments and attacks **should** be taken into consideration.

On a more frequent basis, the actions of the humans that use, operate and maintain the system **should** be reviewed to verify that the humans continue to follow established security procedures. Strong cryptographic systems can be compromised by lax and inappropriate human actions. Highly unusual events **should** be noted and reviewed as possible indicators of attempted attacks on the system.

9.3 Key Management System Survivability

9.3.1 Backup Keys

[OMB11/01] notes that encryption is an important tool for protecting the confidentiality of disclosure-sensitive information that is entrusted to an agency's care, but that the encryption of agency data also presents risks to the availability of information needed for mission performance. Agencies are reminded of the need to protect the continuity of their information technology operations and agency services when implementing encryption. The guidance specifically notes that, without access to the cryptographic keys that are needed to decrypt information, organizations risk the loss of their access to that information. Consequently, it is prudent to retain backup copies of the keys necessary to decrypt stored enciphered information, including master keys, key-encrypting keys, and the related keying material necessary to decrypt encrypted information until there is no longer any requirement for access to the underlying plaintext information (see Tables 7 and 8 in Section 8.2.2.1).

As the tables show, there are other operational keys in addition to those associated with decryption that organizations may need to backup (e.g. public signature-verification keys and authorization keys). Backup copies of keying material **shall** be stored in accordance with the provisions of Section 6 in order to protect the confidentiality of encrypted information and the integrity of source authentication, data integrity, and authorization processes.

9.3.2 Key Recovery

There are a number of issues associated with key recovery. An extensive discussion is provided in Appendix B. Key recovery issues to be addressed include:

1. Which keying material, if any, needs to be backed up or archived for later recovery?

2. Where will backed-up or archived keying material be stored?

3. When will archiving be done (e.g., during key activation or at the end of a key's cryptoperiod)?

4. Who will be responsible for protecting the backed-up or archived keying material?

5. What procedures need to be put in place for storing and recovering the keying material?

6. Who can request a recovery of the keying material and under what conditions?

7. Who will be notified when a key recovery has taken place and under what conditions?

8. What audit or accounting functions need to be performed to ensure that the keying material is only provided to authorized entities?

9.3.3 System Redundancy/Contingency Planning

Cryptography is a useful tool for preventing unauthorized access to data and/or resources, but when the mechanism fails, it can prevent access by valid users to critical information and processes. Loss or corruption of the only copy of cryptographic keys can deny users access to information. For example, a locksmith can usually defeat a broken physical mechanism, but access to information encrypted by a strong algorithm may not be practical without the correct decryption key. The continuity of an organization's operations can depend heavily on contingency planning for key management systems that includes a redundancy of critical logical processes and elements, including key management and cryptographic keys.

9.3.3.1 General Principles

Planning for recovery from system failures is an essential management function. Interruptions of critical infrastructure services **should** be anticipated, and planning for maintaining the continuity of operations in support of an organization's primary mission requirements **should** be done. With respect to key management, the following situations are typical of those for which planning is necessary:

1. Lost key cards or tokens,

2. Forgotten passwords that control access to keys,

3. Failure of key input devices (e.g., readers),

4. Loss or corruption of the memory media on which keys and/or certificates are stored,

5. Compromise of keys,

6. Corruption of Certificate Revocation Lists (CRLs) or Compromised Key Lists (CKLs),

7. Hardware failure of key or certificate generation, registration, and/or distribution systems, subsystems, or components,

8. Power loss requiring re-initialization of key or certificate generation, registration, and/or distribution systems, subsystems, or components,

9. Corruption of the memory media necessary for key or certificate generation, registration, and/or distribution systems, subsystems, or components,

10. Corruption or loss of key or certificate distribution records and/or audit logs,

11. Loss or corruption of the association of keying material to the holders/users of the keying material, and

12. Unavailability of older software or hardware that is needed to access keying material or process protected information.

While recovery discussions most commonly focus on the recovery of encrypted data and the restoration of encrypted communications capabilities, planning **should** also address 1) the restoration of access (without creating temporary loss of access protections) where cryptography is used in access control mechanisms, 2) the restoration of critical processes (without creating temporary loss of privilege restrictions) where cryptography is used in authorization mechanisms, and 3) the maintenance/restoration of integrity protection in digital signature and message authentication applications.

Contingency planning **should** include 1) providing a means and assigning responsibilities for rapidly recognizing and reporting critical failures; 2) the assignment of responsibilities and the placement of resources for bypassing or replacing failed systems, subsystems, and components; and 3) the establishment of detailed bypass and/or recovery procedures.

Contingency planning includes a full range of integrated logistics support functions. Spare parts (including copies of critical software programs, manuals, and data files) **should** be available (acquired or arranged for) and pre-positioned (or delivery-staged). Emergency maintenance, replacement, and/or bypass instructions **should** be prepared and disseminated to both designated individuals and to an accessible and advertised access point. Designated individuals **should** be trained in their assigned recovery procedures, and all personnel **should** be trained in reporting procedures and workstation-specific recovery procedures.

9.3.3.2   Cryptography and Key Management-specific Recovery Issues

Cryptographic keys are relatively small components or data elements that often control access to large volumes of information or critical processes. As the Office of Management and Budget has noted [OMB11/01], "without access to the cryptographic key(s) needed to decrypt information, [an] agency risks losing access to its valuable information." Agencies are reminded of the need to protect the continuity of their information technology operations and agency services when implementing encryption. The guidance particularly stresses that agencies must address information availability and assurance requirements through appropriate data recovery mechanisms, such as cryptographic key recovery.

Key recovery generally involves some redundancy, or multiple copies of keying material. If one copy of a critical key is lost or corrupted, another copy usually needs to be available in order to recover data and/or restore capabilities. At the same time, the more copies of a key that exist and are distributed to different locations, the more susceptible the key usually is to compromise through penetration of the storage location or subversion of the custodian (e.g., user, service agent, key production/distribution facility). In this sense, key confidentiality requirements conflict with continuity of operations requirements. Special care needs to be taken to safeguard all copies of keying material, especially symmetric keys and private (asymmetric) keys. More detail regarding contingency plans and planning requirements is provided in Part 2 of this *Recommendation for Key Management*.

9.3.4 Compromise Recovery

When keying material that is used to protect sensitive information or critical processes is disclosed to unauthorized entities, all of the information and/or processes protected by that keying material becomes immediately subject to disclosure, modification, subversion, and/or denial of service. All compromised keys **shall** be revoked; all affected keys **shall** be replaced; and, where sensitive or critical information or processes are affected, an immediate damage assessment **should** be conducted. Measures necessary to mitigate the consequences of suspected unauthorized access to protected data or processes and to reduce the probability or frequency of future compromises may follow.

Where symmetric keys or private (asymmetric) keys are used to protect only a single user's local information or communications between a single pair of users, the compromise recovery process can be relatively simple and inexpensive. Damage assessment and mitigation measures are often local matters.

On the other hand, where a key is shared by or affects a large number of users, damage can be widespread, and recovery is both complex and expensive. Some examples of keys, the compromise of which might be particularly difficult or expensive to recover from, include the following:

1. A CA's private signature key, especially if it is used to sign a root certificate in a public-key infrastructure

2. Symmetric key-transport key shared by a large number of users

3. Private asymmetric key-transport key shared by a large number of users

4. Master key used in the generation of keys by a large number of users

5. Symmetric data-encryption key used to encrypt data in a large distributed database

6. Symmetric key shared by a large number of communications network participants

7. Key used to protect a large number of stored keys

In all of these cases, a large number of key owners or relying parties (e.g., all parties authorized to use the secret key of a symmetric-key algorithm or the public key of an asymmetric-key algorithm) would need to be immediately notified of the compromise. The inclusion of the key identifier on a Compromised Key List (CKL) or the certificate serial number on a Certificate Revocation List (CRL) to be published at a later date might not be sufficient. This means that a list of (the most-likely) affected entities might need to be maintained, and a means for communicating news of the compromise would be required. Particularly in the case of the compromise of a symmetric key, news of the compromise and the replacement of keys **should** be sent only to the affected entities so as not to encourage others to exploit the situation.

In all of these cases, a secure path for replacing the compromised keys is required. In order to permit rapid restoration of service, an automated (e.g., over-the-air) replacement path is preferred (see Section 8.2.3). In some cases, however, there may be no practical alternative to manual distribution (e.g., compromise of a root CA's private key). Contingency distribution of alternate keys may help restore service rapidly in some circumstances (e.g., compromise of a widely held symmetric key), but the possibility of simultaneous compromise of operational and contingency keys would need to be considered.

Damage assessment can be extraordinarily complex, particularly in cases such as the compromise and replacement of CA private keys, widely used transport keys, and keys used by many users of large distributed databases.

10 Key Management Specifications for Cryptographic Devices or Applications

Key management is often an afterthought in the cryptographic development process. As a result, cryptographic subsystems often fail to support the key management functionality and protocols that are necessary to provide adequate security with the minimum necessary reduction in operational efficiency. All cryptographic development activities **should** involve key management planning and specification (see Part 2) by those managers responsible for the secure implementation of cryptography into an information system. Key management planning **should** begin during the initial conceptual/development stages of the cryptographic development lifecycle, or during the initial discussion stages for the application of existing cryptographic components into information systems and networks. The specifications that result from the planning activities **shall** be consistent with NIST key management guidance.

For cryptographic development efforts, a key specification and acquisition planning process **should** begin as soon as the candidate algorithm(s) and, if appropriate, keying material media and format have been identified. Key management considerations may affect algorithm choice, due to operational efficiency considerations for anticipated applications. For the application of existing cryptographic products for which no key management specification exists, the planning and specification processes **should** begin during device and source selection, and continue through acquisition and installation.

The types of key management components that are required for a specific cryptographic device and/or for suites of devices used by organizations **should** be standardized to the maximum possible extent, and new cryptographic device-development efforts **shall** comply with NIST key-management recommendations. Accordingly, NIST criteria for the security, accuracy, and utility of key management components in electronic and physical forms **shall** be met. Where the criteria for security, accuracy, and utility can be satisfied with standard key management components (e.g., PKI), the use of those compliant components is encouraged. A developer may choose to employ non-compliant key management as a result of security, accuracy, utility, or cost considerations. However, such developments **should** conform as closely as possible to established key management recommendations.

10.1 Key Management Specification Description/Purpose

The Key Management Specification is the document that describes the key management components that may be required to operate a cryptographic device throughout its lifetime. Where applicable, the Key Management Specification also describes key management components that are provided by a cryptographic device. The Key Management Specification documents the capabilities that the cryptographic application requires from key sources (e.g., the Key Management Infrastructure (KMI) described in Part 2 of this *Recommendation for Key Management*).

10.2 Content of the Key Management Specification

The level of detail required for each section of the Key Management Specification can be tailored, depending upon the complexity of the device or application for which the Key Management Specification is being written. The Key Management Specification **should** contain a title page that includes the device identifier, and the developer's or integrator's identifier. A revision page, a list of reference documents, a table of contents, and a definition of abbreviations and acronyms page **should** also be included. The terminology used in a Key Management Specification **shall** be in accordance with the terms defined in appropriate NIST Standards and guidelines. Unless the information is tightly controlled, the Key Management Specification **should not** contain proprietary or sensitive information. [Note: If the cryptographic application is supported by a PKI, a statement to that effect **should** be included in the appropriate Key Management Specification sections below.]

10.2.1 Cryptographic Application

A Cryptographic Application section provides a basis for the development of the rest of the Key Management Specification. The Cryptographic Application section provides a brief description of the cryptographic application or proposed employment of the cryptographic device. This includes the purpose or use of the cryptographic device (or application of a cryptographic device), and whether it is a new cryptographic device, a modification of an existing cryptographic device, or an existing cryptographic device for which a Key Management Specification does not exist. A brief description of the security services (confidentiality, integrity, non-repudiation, access control, identification and authentication, and availability) that the cryptographic device/application provides **should** be included. Information concerning long-term and potential interim key management support (key management components) for the cryptographic application **should** be provided.

10.2.2 Communications Environment

A Communications Environment section provides a brief description of the communications environment in which the cryptographic device is designed to operate. Some examples of communications environments include:

1. Data networks (intranet, Internet, VPN),

2. Wired communications (landline, dedicated or shared switching resources), and

3. Wireless communications (satellite, radio frequency).

The environment may also include any anticipated access controls on communications resources, data sensitivity, privacy issues, non-repudiation requirements, etc.

10.2.3 Key Management Component Requirements

A Key Management Component Requirements section describes the types and logical structure of the keying material required for the operation of the cryptographic device. Cryptographic applications using public-key certificates (i.e., X.509 certificates) **should** describe the types of certificates supported. The following information **should** be included:

1. The different keying material classes or types required, supported, and/or generated (e.g., for PKI: CA, signature, key establishment, and authentication);

2. The key management algorithm(s) (the applicable **approved** algorithms);

3. The keying material format(s) (reference any existing key specification, if known);

4. The set of acceptable PKI policies (as applicable);

5. The tokens to be used.

The description of the key management component format may reference an existing cryptographic device key specification. If the format of the key management components is not already specified, then the format and medium **should** be specified in the Key Management Specification.

10.2.4   Key Management Component Generation

The Key Management Specification **should** include a description of the requirements for the generation of key management components by the cryptographic device for which the Key Management Specification is written. If the cryptographic device does not provide generation capabilities, the key management components that will be required from external sources **should** be identified.

10.2.5   Key Management Component Distribution

Where a device supports the automated distribution of keying material, the Key Management Specification **should** include a description of the distribution and transport encapsulation method(s) (where employed) used for keying material supported by the device. The distribution plan may describe the circumstances under which the key management components are encrypted or unencrypted, their physical form (electronic, paper, etc.), and how they are identified during the distribution process. In the case of a dependence on manual distribution, the dependence and any handling assumptions regarding keying material **should** be stated.

10.2.6   Keying Material Storage

The Key Management Specification **should** address how the cryptographic device or application for which the Key Management Specification is being written stores information, and how the keying material is identified during its storage life (e.g., Distinguished Name). The storage capacity capabilities for information **should** be included.

10.2.7   Access Control

The Key Management Specification **should** address how access to the cryptographic device components and functions is to be authorized, controlled, and validated to request, generate, handle, distribute, store, and/or use keying material. Any use of passwords and personal identification numbers (PINs) **should** be included. For PKI cryptographic applications, role-based privileging and the use of any tokens **should** be described.

10.2.8   Accounting

The Key Management Specification **should** describe any device or application support for accounting of the keying material. Any support for or outputs to logs used to support the tracking of key management component generation, distribution, storage, use and/or destruction **should** be detailed. The use of appropriate privileging to support the control of keying material that is used by the cryptographic application **should** also be described, in addition to the directory capabilities used to support PKI cryptographic applications, if applicable. The Key Management Specification **shall** identify where human and automated tracking actions are required and where multi-party control is required, if applicable. Section 9.1 of this Recommendation provides accountability guidance.

10.2.9   Compromise Management and Recovery

The Key Management Specification **should** address any support for the restoration of protected communications in the event of the compromise of keying material used by the cryptographic device/application. The recovery process description **should** include the methods for re-keying. For PKI cryptographic applications, the implementation of Certificate Revocation Lists (CRLs) and Compromised Key Lists (CKLs) **should** be detailed. For system specifications, a description of how certificates will be reissued and renewed within the cryptographic application **should** also be included. General compromise recovery guidance is provided in Section 9.3.4 of this Recommendation.

10.2.10 Key Recovery

The Key Management Specification **should** include a description of product support or system mechanisms for effecting key recovery. Key recovery addresses how unavailable encryption keys can be recovered. System developers **should** include a discussion of the generation, storage, and access to long-term storage keys in the key-recovery-process description. The process of transitioning from the current to future long-term storage keys **should** also be described. General contingency planning guidance is provided in Section 9.3.3 of this Recommendation. Key recovery is treated in detail in Appendix B, Key Recovery.

APPENDIX A: Cryptographic and Non-cryptographic Integrity and Authentication Mechanisms

Integrity and authentication services are particularly important in protocols that include key management. When integrity or authentication services are discussed in this Recommendation, they are afforded by "strong" cryptographic integrity or authentication mechanisms. Secure communications and key management are typically provided using a communications protocol that offers certain services, such as integrity protection or a "reliable" transport service. However, the integrity protection or reliable transport services of communications protocols are not necessarily adequate for cryptographic applications, particularly for key management, and there might be confusion about the meaning of terms such as "integrity".

All communications channels have some noise (i.e., unintentional errors inserted by the transmission media), and other factors, such as network congestion, can cause packets to be lost. Therefore, integrity protection and reliable transport services for communications protocols are designed to function over a channel with certain worst-case noise characteristics. Transmission bit errors are typically detected using 1) a non-cryptographic checksum[31] to detect transmission errors in a packet, and 2) a packet counter that is used to detect lost packets. A receiving entity that detects damaged packets (i.e., packets that contain bit errors) or lost packets may request the sender to retransmit them. The non-cryptographic checksums are generally effective at detecting transmission noise. For example, the common CRC-32 checksum algorithm used in local area network applications detects all error bursts with a span of less than 32 bits, and detects longer random bursts with a 2^{-32} failure probability. However, the non-cryptographic CRC-32 checksum does not detect the swapping of 32-bit message words, and specific errors in particular message bits cause predictable changes in the CRC-32 checksum. The sophisticated attacker can take advantage of this to create altered messages that pass the CRC-32 integrity checks, even, in some cases, when the message is encrypted.

Forward error-correcting codes are a subset of non-cryptographic checksums that can be used to correct a limited number of errors without retransmission. These codes may be used as checksums, depending on the application and noise properties of the channel.

Cryptographic integrity mechanisms, on the other hand, protect against an active, intelligent attacker who might attempt to disguise his attack as noise. Typically, the bits altered by the attacker are not random; they are targeted at system properties and vulnerabilities. Cryptographic integrity mechanisms are effective in detecting random noise events, but they also detect the more systematic deliberate attacks. Cryptographic hash functions, such as SHA-256 are designed to make every bit of the hash value a complex, nonlinear function of every bit of the message text, and to make it impractical to find two messages that hash to the same value. On average, it is necessary to perform 2^{128} SHA-256 hash operations to find two messages that hash to the same value, and it is much harder to find another message whose SHA-256 hash is the same value as the hash of any given message. Cryptographic message authentication code (MAC) algorithms employ hash functions or symmetric encryption algorithms and a key to authenticate the source

---

[31] Checksum: an algorithm that uses the bits in the transmission to create a checksum value. The checksum value is normally sent in the transmission. The receiver re-computes the checksum value using the bits in the received transmission, and compares the received checksum value with the computed value to determine whether or not the transmission was correctly received. A non-cryptographic checksum algorithm uses a well-known algorithm without secret information (i.e., a cryptographic key).

of a message, as well as protect its integrity (i.e., detect errors). Digital signatures use public-key algorithms and hash functions to provide both authentication and integrity services. Compared to non-cryptographic integrity or authentication mechanisms, these cryptographic services are usually computationally more expensive; this seems to be unavoidable, since cryptographic protections must also resist deliberate attacks by knowledgeable adversaries with substantial resources.

Cryptographic and non-cryptographic integrity mechanisms may be used together. For example, consider the TLS protocol (see Part 3). In TLS, a client and a server can authenticate each other, establish a shared "master key" and transfer encrypted payload data. Every step in the entire TLS protocol run is protected by cryptographic integrity and authentication mechanisms, and the payload is usually encrypted. Like most cryptographic protocols, TLS will detect any attack or noise event that alters any part of the protocol run with a given probability. However, TLS has no error recovery protocol. If an error is detected, the protocol run is simply terminated. Starting a new TLS protocol run is quite expensive. Therefore, TLS requires a "reliable" transport service, typically the Internet Transport Control Protocol (TCP), to handle and recover from ordinary network transmission errors. TLS will detect errors caused by an attack or noise event, but has no mechanism to recover from them. TCP will generally detect such errors on a packet-by-packet basis and recover from them by retransmission of individual packets, before delivering the data to TLS. Both TLS and TCP have integrity mechanisms, but a sophisticated attacker could easily fool the weaker non-cryptographic checksums of TCP. However, because of the cryptographic integrity mechanism provided in TLS, the attack is thwarted.

There are some interactions between cryptographic and non-cryptographic integrity or error-correction mechanisms that users and protocol designers must take into account. For example, many encryption modes expand ciphertext errors: a single bit error in the ciphertext can change an entire block or more of the resulting plaintext. If forward error correction is applied before encryption, and errors are inserted in the ciphertext during transmission, the error expansion during the decryption might "overwhelm" the error correction mechanism, making the errors uncorrectable. Therefore, it is preferable to apply the forward error-correction mechanism after the encryption process. This will allow the correction of errors by the receiving entity's system before the ciphertext is decrypted, resulting in "correct" plaintext.

Interactions between cryptographic and non-cryptographic mechanisms can also result in security vulnerabilities. One classic way this occurs is with protocols that use stream ciphers[32] with non-cryptographic checksums (e.g. CRC-32) that are computed over the plaintext data and that acknowledge good packets. An attacker can copy the encrypted packet, selectively modify individual ciphertext bits, selectively change bits in the CRC, and then send the packet. Using the protocol's acknowledgement mechanism, the attacker can determine when the CRC is correct, and therefore, determine certain bits of the underlying plaintext. At least one widely used wireless wireless-encryption protocol has been broken with such an attack.

---

[32] Stream ciphers encrypt and decrypt one element (e.g., bit or byte) at a time. There are no **approved** algorithms specifically designated as stream ciphers. However, some of the cryptographic modes defined in [SP 800-38] can be used with a symmetric block cipher algorithm, such as AES, to perform the function of a stream cipher.

APPENDIX B: Key Recovery

Federal agencies have a responsibility to protect the information contained in, processed by and transmitted between their information technology systems. Cryptographic techniques are often used as part of this process. These techniques are used to provide confidentiality, assurance of integrity, non-repudiation or access control. Policies **shall** be established to address the protection and continued accessibility of cryptographically protected information, and procedures **shall** be in place to ensure that the information remains viable during its lifetime. When cryptographic keying material is used to protect the information, this same keying material may need to be available to remove (e.g., decrypt) or verify (e.g., verify the MAC) those protections.

In many cases, the keying material used for cryptographic processes might not be readily available. This might be the case for a number of reasons, including:

1. The cryptoperiod of the key has expired, and the keying material is no longer in operational storage,

2. The keying material has been corrupted (e.g., the system has crashed or a virus has modified the saved keying material in operational storage), or

3. The owner of the keying material is not available, and the owner's organization needs to obtain the plaintext information.

In order to have this keying material available when required, the keying material needs to be saved somewhere or to be constructible (e.g., derivable) from other available keying material. The process of re-acquiring the keying material is called key recovery. Key recovery is often used as one method of information recovery when the plaintext information needs to be recovered from encrypted information. However, keying material or other related information may need to be recovered for other reasons, such as the corruption of keying material in normal operational storage (see Section 8.2.1), e.g., the verification of MACs for archived documents. Key recovery may also be appropriate for situations in which it is easier or faster to recover the keying material than it is to generate and distribute new keying material.

However, there are applications that may not need to save the keying material for an extended time because of other procedures to recover an operational capability when the keying material or the information protected by the keying material becomes inaccessible. Applications of this type could include telecommunications where the transmitted information could be resent, or applications that could quickly derive, or acquire and distribute new keying material.

It is the responsibility of an organization to determine whether or not the recovery of keying material is required for their application. The decision as to whether key recovery is required **should** be made on a case-by-case basis, and this decision **should** be reflected in the Key Management Policy and the Key Management Practices Statement (see Part 2). If the decision is made to provide key recovery, the appropriate method of key recovery **should** be selected, designed and implemented, based on the type of keying material to be recovered; an appropriate entity needs to be selected to maintain the backup or archive database and manage the key recovery process.

If the decision is made to provide key recovery for a key, all information associated with that key **shall** also be recoverable (see Table 5 in Section 6).

B.1 Recovery from Stored Keying Material

The primary purpose of backing up or archiving keying material is to be able to recover that material when it is not otherwise available. For example, encrypted information cannot be transformed into plaintext information if the decryption key is lost or modified; the integrity of data cannot be determined if the key used to verify the integrity of that data is not available. The key recovery process retrieves the keying material from backup or archive storage, and places it either in a device or module, or in other immediately accessible storage (see Section 8.3.1).

B.2 Recovery by Reconstruction of Keying Material

Some keying material may be recovered by reconstructing or re-deriving the keying material from other available keying material, the "base" keying material (e.g., a master key for a key-derivation method). The base keying material **shall** be available in normal operational storage (see Section 8.2.1), backup storage (see Section 8.2.2.1) or archive storage (see Section 8.3.1).

B.3 Conditions Under Which Keying Material Needs to be Recoverable

The decision as to whether to back up or archive keying material for possible key recovery **should** be made on a case-by-case basis. The decision **should** be based on:

1. The type of key (e.g., signing private key, long-term data-encryption key),

2. The application in which the key will be used (e.g., interactive communications, file storage),

3. Whether the key is "owned" by the local entity (e.g., a private key) or by another entity (e.g., the other entity's public key) or is shared (e.g., a symmetric data-encryption key shared by two entities),

4. The role of the entity in a communication (e.g., sender or receiver),

5. The algorithm or computation in which the key will be used (e.g., does the entity have the necessary information to perform a given computation if the key were to be recovered), and

6. The value of the information protected by the keying material, and the consequences of the loss of the keying material.

The factors involved in a decision for or against key recovery **should** be carefully assessed. The trade-offs include continuity of operations, versus the risk of possibly exposing the keying material and the information it protects if control of the keying material is lost.

When the key-recovery operation is requested by the key's owner, the following actions **shall** be taken:

1. If the key is lost with the possibility of having been compromised, then the key **shall** be replaced as soon as possible after recovery in order to limit the exposure of the recovered key and the data it protects (see Section 8.2.3.1). This requires reapplying the protection on the protected data using the new key. For example, suppose that the key (Key_A) that was used to encrypt data has been misplaced in a manner in which it could have been compromised. As soon as possible after Key_A is recovered, Key_A **shall** be used to decrypt the data, and the data **shall** be re-encrypted under a new key (Key_B). Key_B **shall** have no relationship to Key_A (e.g., Key_B **shall not** be an update of Key_A).

2. If the key becomes inaccessible or has been modified, but compromise is not suspected, then the key may be recovered. No further action is required (e.g., re-encrypting the data). For example, if the key becomes inaccessible because the system containing the key crashes, or the key is inadvertently overwritten and a compromise is not suspected, then the key may simply be restored.

The following subsections provide discussions to assist an organization in determining whether or not key recovery is needed. Although the following discussions address only the recoverability of keys, any related information **shall** also be recoverable.

B.3.1 Signature Key Pairs

The private key of a signature key pair (the private signature key) is used by the owner of the key pair to apply digital signatures to information. The associated public key (the public signature-verification key) is used by relying entities to verify the digital signature.

B.3.1.1 Private Signature Keys

Private signature keys **shall not** be archived (see Table 9 in Section 8.3.1). Key backup is not usually desirable for the private key of a signing key pair, since the non-repudiability of the signature comes into question. However, exceptions may exist. For example, replacing the private signature key and having its associated public signature-verification key distributed (in accordance with Section 8.1.5.1) in a timely manner may not be possible under some circumstances, so recovering the private signature key from backup storage may be justified. This may be the case, for example, for the private signature key of a CA. If a private signature key is backed up, the private signature key **shall** be recovered using a highly secure method. Depending on circumstances, the key **should** be recovered for immediate use only, and then **shall** be replaced as soon after the recovery process as possible.

Instead of backing up the private signature key, a second private signature key and associated public key could be generated, and the public key distributed in accordance with Section 8.1.5.1 for use if the primary private signature key becomes unavailable. If backup is considered for the private signature key, an assessment **should** be made as to its importance and the time needed to recover the key, as opposed to the time needed to generate a new key pair, and certify and distribute a new public signature-verification key.

A private signature key **shall not** be archived.

B.3.1.2 Public Signature-verification Keys

It is appropriate to backup or archive a public signature-verification key for as long as required in order to verify the information signed by the associated private signature key. In the case of a public key that has been certified (e.g., by a Certification Authority), saving the public-key certificate would be an appropriate form of storing the public key; backup or archive storage may be provided by the infrastructure (e.g., by a certificate repository). The public key **should** be stored in backup storage until the end of the private key's cryptoperiod, and **should** be stored in archive storage as long as required for the verification of signed data.

B.3.2 Symmetric Authentication Keys

A symmetric authentication key is used to provide assurance of the integrity and source of information. A symmetric authentication key can be used:

1. By an originator to create a message authentication code (MAC) that can be verified at a later time to determine the authenticity or integrity of the authenticated information; the authenticated information and its MAC could then be stored for later retrieval or transmitted to another entity,

2. By an entity that retrieves the authenticated information and the MAC from storage to determine the integrity of the stored information (Note: This is not a communication application),

3. Immediately upon receipt by a receiving entity to determine the integrity of transmitted information and the source of that information (the received MAC and the associated authenticated information may or may not be subsequently stored), or

4. By a receiving and retrieving entity to determine the integrity and source of information that has been received and subsequently stored using the same MAC (and the same authentication key); checking the MAC is not performed prior to storage.

For each of the above cases, a decision to provide a key recovery capability **should** be made, based on the following considerations.

In case 1, the symmetric authentication key need not be backed up or archived if the originator can establish a new authentication key prior to computing the MAC, making the key available to any entity that would need to subsequently verify the information that is authenticated using this new key. If a new authentication key cannot be obtained in a timely manner, then the authentication key **should** be backed up or archived.

In case 2, the symmetric authentication key **should** be backed up or archived for as long as the integrity of the information needs to be determined.

In case 3, the symmetric authentication key need not be backed up or archived if the authentication key can be resent to the recipient. In this case, establishing and distributing a new symmetric authentication key, rather than reusing the "lost" key, is also acceptable; a new MAC would need to be computed on the information using the new authentication key. Otherwise, the symmetric authentication key **should** be backed up. Archiving the authentication key is not appropriate if the MAC and the authenticated information are not subsequently stored, since the use of the key for both applying and checking the MAC would be discontinued at the end of the key's cryptoperiod. If the MAC and the authenticated information are subsequently stored, then the symmetric authentication key **should** be backed up or archived for as long as the integrity and source of the information needs to be determined.

In case 4, the symmetric authentication key **should** be backed up or archived for as long as the integrity and source of the information needs to be determined.

The symmetric authentication key may be stored in backup storage for the cryptoperiod of the key, and in archive storage until no longer required. If the authentication key is recovered by reconstruction, the "base" key (e.g., the master key for a key-derivation method) may be stored in normal operational storage or backup storage for the cryptoperiod of the key, and in archive storage until no longer required.

B.3.3   Authentication Key Pairs

A public authentication key is used by a receiving entity to obtain assurance of the identity of the originating entity when executing an authentication mechanism. The associated private authentication key is used by the originating entity to provide this assurance to a receiving entity by computing a digital signature on the information. This key pair may not provide non-repudiation.

B.3.3.1   Public Authentication Keys

It is appropriate to store a public authentication key in either backup or archive storage for as long as required to verify the authenticity of the data that was authenticated by the associated private authentication key.

In the case of a public key that has been certified (e.g., by a Certification Authority), saving the public-key certificate would be an appropriate form of storing the public key; backup or archive storage may be provided by the infrastructure (e.g., by a certificate repository). The public key may be stored in backup storage until the end of the private key's cryptoperiod, and may be stored in archive storage as long as required.

B.3.3.2   Private Authentication Keys

When the private key is used only for the authentication of transmitted data, whether or not the authenticated data is subsequently stored, the private authentication key need not be backed up if a new key pair can be generated and distributed in accordance with Section 8.1.5.1 in a timely manner. However, if a new key pair cannot be generated quickly, the private key **should** be stored in backup storage during the cryptoperiod of the private key. The private key **shall not** be stored in archive storage.

When the private authentication key is used to protect stored information only, the private authentication key **should not** be backed up if a new key pair can be generated. However, if a new key pair cannot be generated, the private key **should** be stored in backup storage during the cryptoperiod of the private key. The private key **shall not** be stored in archive storage.

B.3.4   Symmetric Data-Encryption Keys

A symmetric data-encryption key is used to protect the confidentiality of stored or transmitted information or both. The same key is used initially to encrypt the plaintext information to be protected, and later to decrypt the encrypted information (i.e., the ciphertext), thus obtaining the original plaintext.

The key needs to be available for as long as any information that is encrypted using that key may need to be decrypted. Therefore, the key **should** be backed up or archived during this period. However, at some time, the strength of the cryptographic protection could be reduced or lost completely; for example, the encryption algorithm may no longer offer adequate security, or the symmetric key may have been compromised. If the encryption algorithm or the key no longer provide the required security (e.g., the length of the key is no longer considered adequate, or the key has been compromised), then the cryptographic protection **shall** be regarded as inadequate. Appropriate storage systems are being developed that employ cryptographic timestamps to store sensitive data beyond the security life of the encryption algorithm or the data-encryption key (e.g., to provide assurance about the date of the encryption process, so that it can be determined whether the algorithm and key provided sufficient protection at that time, as well as to provide assurance that the encrypted data has been physically protected from compromise).

In order to allow key recovery, the symmetric data-encryption key **should** be stored in backup storage during the cryptoperiod of the key, and **should** be stored in archive storage, if required. In many cases, the key is protected and stored with the encrypted data. The key is wrapped (i.e., encrypted) by an archive-encryption key or by a symmetric key-wrapping key that is wrapped by a protected archive-encryption key.

A symmetric-data encryption key that is used only for transmission is used by an originating entity to encrypt information, and by the receiving entity to decrypt the information immediately upon receipt. If the data-encryption key is lost or corrupted, and a new data-encryption key can be easily obtained by the originating and receiving entities, then the key need not be backed up. However, if the key cannot be easily replaced by a new key, then the key **should** be backed up if the information to be exchanged is of sufficient importance. The data-encryption key may not need to be archived when used for transmission only.

B.3.5 Symmetric Key-Wrapping Keys

A symmetric key-wrapping key is used to wrap (i.e., encrypt) keying material that is to be protected, and may be used to protect multiple sets of keying material. The protected keying material is then transmitted or stored or both.

If a symmetric key-wrapping key is used only to transmit keying material, and the key-wrapping key becomes unavailable (e.g., is lost or corrupted), it may be possible to either resend the key-wrapping key, or to establish a new key-wrapping key and use it to resend the keying material. If this is possible within a reasonable timeframe, backup of the key-wrapping key is not necessary. If the key-wrapping key cannot be resent, or a new key-wrapping key cannot be readily obtained, backup of the key-wrapping key **should** be considered. The archive of a key-wrapping key that is only used to transmit keying material may not be necessary.

If a symmetric key-wrapping key is used to protect keying material in storage, then the key-wrapping key **should** be backed up or archived for as long as the protected keying material may need to be accessed. However, at some time, the strength of the key-wrapping mechanism may be reduced or lost completely; for example, the key-wrapping algorithm may no longer offer adequate security, or the key-wrapping key may have been compromised. If the wrapping algorithm or the key-wrapping key no longer provide the required security (e.g., the length of the key is no longer considered adequate, or the key has been compromised), then the cryptographic protection **shall** be regarded as inadequate. Appropriate storage systems are being developed that employ cryptographic timestamps to store sensitive data beyond the security life of the key-wrapping algorithm or its key-wrapping keys (e.g., to provide assurance about the date of the wrapping process, so that it can be determined whether the key-wrapping algorithm and key-wrapping key provided sufficient protection at that time, as well as to provide assurance that the wrapped keying material data has been physically protected from compromise).

B.3.6 Random Number Generation Keys

A key used for deterministic random bit generation **shall not** be backed up or archived. If this key is lost or modified, it **shall** be replaced with a new key.

B.3.7 Symmetric Master Keys

A symmetric master key is normally used to derive one or more other keys. It **shall not** be used for any other purpose.

The determination as to whether or not a symmetric master key needs to be backed up or archived depends on a number of factors:

1. How easy is it to establish a new symmetric master key? If the master key is distributed manually (e.g., in smart cards or in hard copy by receipted mail), the master key **should** be backed up or archived. If a new master key can be easily and quickly established using automated key-establishment protocols, then the backup or archiving of the master key may not be necessary or desirable, depending on the application.

2. Are the derived keys recoverable without the use of the symmetric master key? If the derived keys do not need to be backed up or archived (e.g., because of their use) or recovery of the derived keys does not depend on reconstruction from the master key (e.g., the derived keys are stored in an encrypted form), then the backup or archiving of the master key may not be desirable. If the derived keys need to be backed up or archived, and the method of key recovery requires reconstruction of the derived key from the master key, then the master key **should** be backed up or archived.

B.3.8 Key-Transport Key Pairs

A key-transport key pair may be used to transport keying material from an originating entity to a receiving entity during communications, or to protect keying material while in storage. The originating entity in a communication (or the entity initiating the storage of the keying material) uses the public key to encrypt the keying material; the receiving entity (or the entity retrieving the stored keying material) uses the private key to decrypt the encrypted keying material.

B.3.8.1 Private Key-Transport Keys

If a key-transport key pair is only used during communications, then the private key-transport key does not need to be backed up if a replacement key pair can be generated and distributed in a timely fashion. Alternatively, one or more additional key pairs could be made available (i.e., already generated and distributed). Otherwise, the private key **should** be backed up. The private key-transport key may be archived.

If the transport key pair is used during storage, then the private key-transport key **should** be backed up or archived for as long as the protected keying material may need to be accessed.

B.3.8.2 Public Key Transport Keys

Backup or archiving of the public key may be done, but may not be necessary.

If the sending entity (the originating entity in a communications) loses the public key-transport key or determines that the key has been corrupted, the key can be reacquired from the key pair owner or by obtaining the public-key certificate containing the public key (if the public key was certified).

If the entity that applies the cryptographic protection to keying material that is to be stored determines that the public key-transport key has been lost or corrupted, the entity may recover in one of the following ways:

1. If the public key has been certified and is stored elsewhere within the infrastructure, then the certificate can be requested.

2. If some other entity knows the public key (e.g., the owner of the key pair), the key can be requested from this other entity.

3. If the private key is known, then the public key can be recomputed.

4. A new key pair can be generated.

B.3.9 Symmetric Key Agreement Keys

Symmetric key-agreement keys are used to establish keying material (e.g., symmetric key-wrapping keys, symmetric data-encryption keys, or symmetric authentication keys). Each key-agreement key is shared between two or more entities. If these keys are distributed manually (e.g., in a key loading device or by receipted mail), then the symmetric key-agreement key **should** be backed up. If an automated means is available for quickly establishing new keys (e.g., a key-transport mechanism can be used to establish a new symmetric key-agreement key), then a symmetric key-agreement key need not be backed up.

Symmetric key-agreement keys may be archived.

B.3.10 Static Key-Agreement Key Pairs

Static key-agreement key pairs are used to establish shared secrets between entities, often in conjunction with ephemeral key pairs (see [SP800-56A] and [SP800-56B]). Each entity uses its private key-agreement key(s), the other entity's public key-agreement key(s) and possibly its public key-agreement key(s) to determine the shared secret. The shared secret is subsequently used to derive shared keying material. Note that in some key-agreement schemes, one or more of the entities may not have a static key-agreement pair (see [SP800-56A] and [SP800-56B]).

B.3.10.1 Private Static Key-Agreement Keys

If the private static key-agreement key cannot be replaced in a timely manner, or if it needs to be retained in order to recover encrypted stored data, then the private key **should** be backed up in order to continue operations. The private key may be archived.

B.3.10.2 Public Static Key Agreement Keys

If an entity determines that the public static key-agreement key is lost or corrupted, the entity may recover in one of the following ways:

1. If the public key has been certified and is stored elsewhere within the infrastructure, then the certificate can be requested.

2. If some other entity knows the public key (e.g., the other entity is the owner of the key pair), the key can be requested from this other entity.

3. If the private key is known, then the public key can be recomputed.

4. If the entity is the owner of the key pair, a new key pair can be generated and distributed.

If none of these alternatives are possible, then the public static key-agreement key **should** be backed up. The public key may be archived.

B.3.11 Ephemeral Key Pairs

Ephemeral key-agreement keys are generated and distributed during a single key-agreement process (e.g., at the beginning of a communication session) and are not reused. These key pairs are used to establish a shared secret (often in combination with static key pairs); the shared secret is subsequently used to derive shared keying material. Not all key-agreement schemes use ephemeral key pairs, and when used, not all entities have an ephemeral key pair (see [SP800-56A]).

B.3.11.1 Private Ephemeral Keys

Private ephemeral keys **shall not**[33] be backed up or archived. If the private ephemeral key is lost or corrupted, a new key pair **shall** be generated, and the new public ephemeral key **shall** be provided to the other participating entity in the key-agreement process.

B.3.11.2 Public Ephemeral Keys

Public ephemeral keys may be backed up or archived. This will allow the reconstruction of the established keying material, as long as the private ephemeral keys are not required in the key-agreement computation.

B.3.12 Symmetric Authorization Keys

Symmetric authorization keys are used to provide privileges to an entity (e.g., access to certain information or authorization to perform certain functions). Loss of these keys will deny the privileges (e.g., prohibit access and disallow performance of these functions). If the authorization key is lost or corrupted and can be replaced in a timely fashion, then the authorization key need not be backed up. A symmetric authorization key **shall not** be archived.

B.3.13 Authorization Key Pairs

Authorization key pairs are used to provide privileges to an entity. The private key is used to establish the "right" to the privilege; the public key is used to determine that the entity actually has the right to the privilege.

B.3.13.1 Private Authorization Keys

Loss of the private authorization key will deny the privileges (e.g., prohibit access and disallow performance of these functions). If the private key is lost or corrupted and can be replaced in a timely fashion, then the private key need not be backed up. Otherwise, the private key **should** be backed up. The private key **shall not** be archived.

B.3.13.2 Public Authorization Keys

If the authorization key pair can be replaced in a timely fashion (i.e., regeneration of the key pair and secure distribution of the private key to the entity seeking authorization), then the public authorization key need not be backed up. Otherwise, the public key **should** be backed up. There is no restriction about archiving the public key.

---

[33] SP 800-56A states that the private ephemeral keys **shall** be destroyed immediately after use. This implies that the private ephemeral keys **shall not** be backed up or archived.

B.3.14 Other Cryptographically Related Material

Like keys, other cryptographically related material may need to be backed up or archived, depending on use.

B.3.14.1 Domain Parameters

Domain parameters are used in conjunction with some public key algorithms to generate key pairs. They are also used with key pairs to create and verify digital signatures, to establish keying material, or to generate random numbers. The same set of domain parameters is often, but not always, used by a large number of entities.

When an entity (entity A) generates new domain parameters, these domain parameters are used in subsequent digital signature generation or key-establishment processes. The domain parameters need to be provided to other entities that need to verify the digital signatures or with whom keys will be established. If the entity (entity A) determines that its copies of the domain parameters have been lost or corrupted, and if the new domain parameters cannot be securely distributed in a timely fashion, then the domain parameters **should** be backed up or archived. Another entity (entity B) **should** backup or archive entity A's domain parameters until no longer required unless the domain parameters can be otherwise obtained (e.g., from entity A).

When the same set of domain parameters are used by multiple entities, the domain parameters **should** be backed up or archived until no longer required unless the domain parameters can be otherwise obtained (e.g., from a trusted source).

B.3.14.2 Initialization Vectors (IVs)

IVs are used by several modes of operation during the encryption or authentication of information using block cipher algorithms. IVs are often stored with the data that they protect. If not, they **should** be backed up or archived as long as the information protected using those IVs needs to be processed (e.g., decrypted or authenticated).

B.3.14.3 Shared Secrets

Shared secrets are generated by each entity participating in a key-agreement process. The shared secret is then used to derive the shared keying material to be used in subsequent cryptographic operations. Shared secrets may be generated during interactive communications (e.g., where both entities are online) or during non-interactive communications (e.g., in store and forward applications).

A shared secret **shall not** be backed up or archived.

B.3.14.4 RNG Seeds

RNG seeds are used in the generation of deterministic random bits that need to remain secret. These seeds **shall not** be shared with other entities. RNG seeds **shall not** be backed up or archived.

B.3.14.5 Other Public and Secret Information

Public and secret information is often used during key establishment. The information may need to be available to determine the keys that are needed to process cryptographically protected information (e.g., to decrypt or authenticate); therefore, the information **should** be backed up or archived until no longer needed to process the protected information.

B.3.14.6   Intermediate Results

The intermediate results of a cryptographic operation **shall not** be backed up or archived.

B.3.14.7   Key Control Information

Key control information is used, for example, to determine the keys and other information to be used to process cryptographically protected information (e.g., decrypt or authenticate), to identify the purpose of a key, or to identify the entities that share the key (see Section 6.2.3). This information is contained in the key's metadata (see Section 6.2.3.1).

Key control information **should** be backed up or archived for as long as the associated key needs to be available.

B.3.14.8   Random Numbers

Random numbers are generated by random number generators. The backup or archiving of a random number depends on how it is used.

B.3.14.9   Passwords

A password is used to acquire access to privileges by an entity, to derive keys or to detect the re-use of passwords.

If the password is only used to acquire access to privileges, and can be replaced in a timely fashion, then the password need not be backed up. In this case, a password **shall not** be archived.

If the password is used to derive cryptographic keys or to prevent the re-use of passwords, the password **should** be backed up and archived.

B.3.14.10   Audit Information

Audit information containing key management events **shall** be backed up and archived.

B.4   Key Recovery Systems

Key recovery is a broad term that may be applied to several different key recovery techniques. Each technique will result in the recovery of a cryptographic key and other information associated with that key (i.e., the keying material). The information required to recover that key may be different for each application or each key recovery technique. The term "Key Recovery Information" (KRI) is used to refer to the aggregate of information that is needed to recover or verify cryptographically protected information. Information that may be considered as KRI includes the keying material to be recovered or sufficient information to reconstruct the keying material, other associated cryptographic information, the time when the key was created, the identifier of the owner of the key (i.e., the individual, application or organization that created the key or that owns the data protected by that key) and any conditions that must be met by a requestor to be able to recover the keying material.

When an organization determines that key recovery is required for all or part of its keying material, a secure Key Recovery System (KRS) needs to be established in accordance with a well-defined Key Recovery Policy (see Appendix B.5). The KRS **shall** support the Key Recovery Policy and consists of the techniques and facilities for saving and recovering the keying material, the procedures for administering the system, and the personnel associated with the system.

When key recovery is determined to be necessary, the KRI may be stored either within an organization (in backup or archive storage) or may be stored at a remote site by a trusted entity. There are many acceptable methods for enabling key recovery. A KRS could be established using a safe for keying material storage; a KRS might use a single computer that provides the initial protection of the plaintext information, storage of the associated keying material and recovery of that keying material; a KRS may include a network of computers with a central Key Recovery Center; or a KRS could be designed using other configurations. Since a KRS provides an alternative means for recovering cryptographic keys, a risk assessment **should** be performed to ensure that the KRS adequately protects the organization's information and reliably provides the KRI when required. It is the responsibility of the organization that needs to provide key recovery to ensure that the Key Recovery Policy, the key recovery methodology, and the Key Recovery System adequately protect the KRI.

A KRS used by the Federal government **shall**:

1. Generate or provide sufficient KRI to allow recovery or verification of protected information.

2. Ensure the validity of the saved key and the other KRI.

3. Ensure that the KRI is stored with persistence and availability that is commensurate with that of the corresponding cryptographically protected data.

4. Use cryptographic modules that are compliant with [FIPS140].

5. Use **approved** algorithms, when cryptography is used.

6. Use algorithms and key lengths that provide security strengths commensurate with the sensitivity of the information associated with the KRI.

7. Be designed to enforce the Key Recovery Policy (see Section B.5).

8. Protect KRI against unauthorized disclosure or destruction. The KRS **shall** verify the source of requests and ensure that only requested and authorized information is provided to the requestor.

9. Protect the KRI from modification.

10. Have the capability of providing an audit trail. The audit trail **shall not** contain the keys that are recovered or any passwords that may be used by the system. The audit trail **should** include the identification of the event being audited, the time of the event, the identifier of the user causing the event, and the success or failure of the event.

11. Limit access to the KRI, the audit trail and authentication data to authorized individuals.

12. Prohibit modification of the audit trail.

B.5   Key Recovery Policy

For each system, application and cryptographic technique used, consideration **shall** be given as to whether or not the keying material may need to be saved for later recovery to allow subsequent decryption or checking the information protected by the keying material. An organization that determines that key recovery is required for some or all of its keying material **should** develop a Key Recovery Policy that addresses the protection and continued accessibility of that information[34] (see [DOD-KRP]). The policy **should** answer the following questions (at a minimum):

1. What keying material needs to be saved for a given application? For example, keys and IVs used for the decryption of stored information may need to be saved. Keys for the authentication of stored or transmitted information may also need to be saved.

2. How and where will the keying material be saved? For example, the keying material could be stored in a safe by the individual who initiates the protection of the information (e.g., the encrypted information), or the keying material could be saved automatically when the protected information is transmitted, received or stored. The keying material could be saved locally or at some remote site.

3. Who will be responsible for protecting the KRI? Each individual, organization or sub-organization could be responsible for their own keying material, or an external organization could perform this function.

4. Who can request key recovery and under what conditions? For example, the individual who protected the information (i.e., used and stored the KRI) or the organization to which the individual is assigned could recover the keying material. Legal requirements may need to be considered. An organization could request the information when the individual who stored the KRI is not available.

5. Under what conditions can the policy be modified and by whom?

6. What audit capabilities and procedures will be included in the KRS? The policy **shall** identify the events to be audited. Auditable events might include KRI requests and their associated responses; who made a request and when; the startup and shutdown of audit functions; the operations performed to read, modify or destroy the audit data; requests to access user authentication data; and the uses of authentication mechanisms.

7. How will the KRS deal with aged keying material[35] or the destruction of the keying material?

8. Who will be notified when keying material is recovered and under what conditions? For example, the individual who encrypted data and stored the KRI could be notified when the organization recovers the decryption key because the person is absent, but the individual might not be notified when the organization is monitoring the activities of that individual.

9. What procedures need to be followed when the KRS or some portion of the data within the KRS is compromised?

---

[34] An organization's key recovery policy may be included in its PKI Certificate Policy.

[35] Keying material whose security strength is now reduced beyond an acceptable level.

APPENDIX C: References

[AC]   Applied Cryptography, Schneier, John Wiley & Sons, 1996.

[ANSX9.31]   Digital Signatures using reversible Public Key Cryptography for the Financial Services Industry (rDSA), 1998.

[ANSX9.44]   Public Key Cryptography for the Financial Services Industry: Key Agreement Using Factoring-Based Cryptography, August 24, 2007.

[ANSX9.62]   Public Key Cryptography for the Financial Services Industry: The Elliptic Curve Digital Signature Algorithm (ECDSA), January 22, 2009.

[DiCrescenzo]   How to forget a secret, G. Di Crescenzo, N. Ferguson, R. Impagliazzo, and M Jakobsson, STACS '99, Available via http://www.macfergus.com/pub/forget.html.

[DOD-KRP]   Key Recovery Policy for the United States Department of Defense, Version 3.0, 31 August 2003, DoD KRP, Attn: I5P, 9800 Savage Road, STE 6737, Ft Meade, MD, 20755-6737.

[FIPS140]   Federal Information Processing Standard 140-2, Security Requirements for Cryptographic Modules, May 25, 2001.

[FIPS180]   Federal Information Processing Standard 180-4, Secure Hash Standard (SHS), March 2012.

[FIPS186]   Federal Information Processing Standard 186-3, Digital Signature Standard (DSS), (Revision of FIPS 186-2, June 2000), June 2009.

[FIPS197]   Federal Information Processing Standard 197, Advanced Encryption Standard (AES), November 2001.

[FIPS198]   Federal Information Processing Standard 198-1, Keyed-Hash Message Authentication Code (HMAC), July 2008.

[FIPS199]   Federal Information Processing Standard 199, Standards for Security Categorization of Federal Information and Information Systems, v 1.0, May 2003.

[HAC]   Handbook of Applied Cryptography, Menezes, van Oorschot and Vanstone, CRC Press, 1996.

[ITLBulletin]   Techniques for System and Data Recovery, NIST ITL Computer Security Bulletin, April 2002.

[OMB11/01]   OMB Guidance to Federal Agencies on Data Availability and Encryption, Office of Management and Budget, November 26, 2001.

[PKCS#1]   PKCS #1 v2.1, RSA Cryptography Standard, RSA Laboratories, June 14, 2002.

[RFC2560]   Request for Comment 2560, X.509 Internet Public Key Infrastructure, Online Certificate Status Protocol – OCSP, IETF Standards Track, June 1999.

[SP800-14]   Special Publication 800-14, Generally Accepted Principles and Practices for Securing Information Technology Systems, September 1996.

[SP800-21]   Special Publication 800-21, Guideline for Implementing Cryptography in the Federal Government, November 1999.

[SP800-32]   Special Publication 800-32, Introduction to Public Key Technology and the Federal PKI Infrastructure, February 2001.

[SP800-37]   Special Publication 800-37, Guide for the Security Certification and Accreditation of Federal Information Systems, May 2004.

[SP800-38]   Special Publication 800-38, Recommendation for Block Cipher Modes of Operation:

SP 800-38A, Methods and Techniques, December 2001.

SP 800-38A (Addendum): Three Variants of Ciphertext Stealing for CBC Mode, October 2010.

SP 800-38B: The CMAC Authentication Mode, May 2005.

SP 800-38C: The CCM Mode for Authentication and Confidentiality, May 2004.

SP 800-38D: Galois/Counter Mode (GCM) and GMAC, November 2007.

SP 800-38E: The XTS-AES Mode for Confidentiality on Storage Devices, January 2010.

SP 800-38F: Recommendation for Block Cipher Modes of Operation: Methods for Key Wrapping, August 2011 (Draft).

[SP800-38A]   Special Publication 800-38A, Recommendation for Block Cipher Modes of Operation-Methods and Techniques, December 2001.

[SP800-38B]   Special Publication 800-38B, Recommendation for Block Cipher Modes of Operation: The CMAC Authentication Mode, May 2005.

[SP800-38F]   Recommendation for Block Cipher Modes of Operation: Methods for Key Wrapping, August 2011 (Draft).

[SP800-56A]   Special Publication 800-56A, Recommendation for Pair-Wise Key Establishment Schemes Using Discrete Logarithm Cryptography, March 2007.

[SP800-56B]   Special Publication 800-56B, Recommendation for Pair-Wise Key Establishment Schemes Using Integer Factorization Cryptography, August 2009.

[SP800-56C]   Special Publication 800-56C, Recommendation for Key Derivation through Extraction-then-Expansion, November 2011.

[SP800-67]   Special Publication 800-67, Recommendation for Triple Data Encryption Algorithm Block Cipher, January 2012.

[SP800-89]   Special Publication 800-89, Recommendation for Obtaining Assurances for Digital Signature Applications, November 2006.

[SP800-90A]   Special Publication 800-90A, Recommendation for Random Number Generation Using Deterministic Random Bit Generators,[36] January 2012.

[SP800-107]   Special Publication 800-107, Recommendation for Applications Using Approved Hash Algorithms, February 2009.

[SP800-108]   Special Publication 800-108, Recommendation for Key Derivation Using Pseudorandom Functions, October 2009.

[SP800-131A]   Special Publication 800-131A, Recommendation for the Transitioning of Cryptographic Algorithms and Key Sizes, January 2011.

[SP800-132]   Special Publication 800-132, Recommendation for Password-Based Key Derivation - Part 1: Storage Applications, December 2010.

[SP800-133]   Special Publication 800-133, Recommendation for Cryptographic Key Generation, August 2011 (Draft).

---

[36] SP 800-90A is a revision of SP 800-90 that was published in March 2007.

APPENDIX D: Revisions

The original version of this document was published in August 2005. In May 2006, the following revisions were incorporated:

1. The definition of security strength has been revised to remove "or security level" from the first column, since this term is not used in the document.

2. In the footnote for 2TDEA in Table 2 of Section 5.6.1, the word "guarantee" has been changed to "assessment".

3. In the paragraph under Table 2 in Section 5.6.1: The change originally identified for the 2006 revision has been superseded by the 2011 revision discussed below.

4. In Table 3 of Section 5.6.1, a list of appropriate hash functions have been inserted into the HMAC and Key Derivation Function columns. In addition, a footnote has been included for the Key Derivation Function column.

5. The original text for the paragraph immediately below Table 3 has been removed.

In March 2007, the following revisions were made to allow the dual use of keys during certificate requests:

1. In Section 5.2, the following text was added:

   "This Recommendation also permits the use of a private key-transport or key-agreement private key to generate a digital signature for the following special case:

   When requesting the (initial) certificate for a static key-establishment key, the associated private key may be used to sign the certificate request. Also refer to Section 8.1.5.1.1.2."

2. In Section 8.1.5.1.1.2, the fourth paragraph was originally as follows:

   "The owner provides POP by performing operations with the private key that satisfy the indicated key use. For example, if a key pair is intended to support key transport, the owner may decrypt a key provided to the owner by the CA that is encrypted using the owner's public key. If the owner can correctly decrypt the ciphertext key using the associated private key and then provide evidence that the key was correctly decrypted (e.g., by encrypting a random challenge from the CA, then the owner has established POP. Where a key pair is intended to support key establishment, POP **shall not** be afforded by generating and verifying a digital signature with the key pair."

   The paragraph was changed to the following, where the changed text is indicated in italics:

   "The *(reputed)* owner **should** *provide* POP by performing operations with the private key that satisfy the indicated key use. For example, if a key pair is intended to support *RSA* key transport, the *CA may provide the owner with a key* that is encrypted using the owner's public key. If the owner can correctly decrypt the ciphertext key using the associated private key and then provide evidence that the key was correctly decrypted (e.g., by encrypting a random challenge from the CA, then the owner has established POP. *However, when a key pair is intended to support key establishment, POP may also be afforded by using the private key to digitally sign the certificate request (although this is not the preferred method). The private key establishment private key (i.e., the private key-agreement or key-transport key)* **shall not** *be used to perform signature operations after certificate issuance.*"

In July 2012, several editorial corrections and clarifications were made, and the following revisions were also made:

1. The Authority section has been updated.

2. Section 1.2: The description of SP800-57, Part 3 has been modified per that document.

3. Section 2.1: Definitions for key-derivation function, key-derivation key, key length, key size, random bit generator and user were added. Definitions for archive, key management archive, key recovery, label, owner, private key, proof of possession, public key, security life of data, seed, shared secret and **should** have been modified. The definition for cryptomodule was removed.

4. Section 2.2: The RBG acronym was inserted.

5. References to FIPS 180-3, FIPS 186-3, SP 800-38, SP 800-56A, SP 800-56B, SP 800-56C, SP 800-89, SP 800-90, SP 800-107, SP 800-108, SP 800-131A, SP 800-132 and SP 800-133 have been corrected or inserted.

6. Section 4.2.4: A footnote was added about the two general types of digital signatures and the focus for this Recommendation.

7. Sections 4.2.5, 4.2.5.3, 4.2.5.5 and 5.3: Discussions about SP 800-56B have been included.

8. Section 5.1.1: The definitions of private signature key, public signature-verification key, symmetric authentication key, private authentication key and public authentication key have been corrected to reflect their current use in systems and protocols. This change is reflected throughout the document.

9. Section 5.1.2, item 3: The description of shared secret has been modified to state that shared secrets are to be protected and handled as if they are cryptographic keys.

10. Sections 5.1.2, 5.3.7, 6.1.2 (Table 5), 8.1.5.3.4, 8.1.5.3.5, 8.2.2.1 (Table 7) and 8.3.1 (Table 9): "Other secret information" has been added to the list of other cryptographic or related information.

11. Section 5.3.1: An additional risk factor was inserted about personnel turnover.

12. Section 5.3.4: A statement was inserted to clarify the difference between the cryptoperiod of a public key and the validity period of a certificate.

13. Section 5.3.6: Statements were inserted that emphasize that longer or shorter cryptoperiods than those suggested may be warranted. Also, further discussion was added about the cryptoperiod of the public ephemeral key-agreement key.

14. Section 5.4.4: A discussion of an owner's assurance of private-key possession was added.

15. Section 5.5: Statements were added about the compromise of a CA's private signature key, and advice was provided for handling such an event.

16. Section 5.6.1: Table 3 and the text preceding the table have been revised for clarity. Additional footnotes were inserted related to table entries, and the footnote about the security strength provided by SHA-1 was modified to indicate that its security strength for digital signature applications remains the subject of speculation.

17. Sections 5.6.2 – 5.6.4: Table 4 and the text preceding it have been modified to be consistent with SP 800-131A. Also, the examples have been modified.

18. Section 5.6.5: This new section was added to address the implications associated with the reduction of security strength because of improvements in computational capabilities or cryptanalysis.

19. Sections 7, 7.1, 7.2 and 7.3: The description of the states and their transitions have been reworded to require specific behavior (e.g., using **shall** or **shall not** statements, rather than containing statement of fact (e.g., using "is" or are").

20. Section 7.3: A discussion of the transition of a private key-transport key and an ephemeral private key-agreement key were added. The previous discussion on private and public key-agreement keys was changed to discuss static private and public key-agreement keys and ephemeral public key-agreement keys.

21. Section 8.1.5.3.4: This section was revised to be more consistent with SP 800-90A.

22. Sections 8.1.5.3.7 and 8.1.5.3.8: New sections were inserted to discuss the distribution of random numbers and passwords.

23. Section 8.1.6: Text was inserted to indicate which keys would or would not be registered.

24. Section 8.2.4: This section was revised to be consistent with SP 800-56A SP 800-56B, SP 800-56C, SP 800-108 and SP 800-132.

25. Section 8.3.1, Table 9: The table was modified to indicate that it is OK to archive the static key-agreement key.

26. Changes were made to Sections 8.3.1; 9.3.2; and Appendices B, B.1, B.3, B.3.1.2, B.3.2, B.3.4, B.3.5, and B.3.10.2 to remove the impression that archiving is only performed after the end of the cryptoperiod of a key (e.g., keys could be archived immediately upon activation), and that the keys in an archive are only of historical interest (e.g., they may be needed to decrypt data long after the cryptoperiod of a key).

27. Section 8.3.3: The discussion about de-registering compromised and non-compromised keys was modified.

28. Section 8.3.5: A discussion about how revocation is achieved for a PKI and for symmetric-key systems was added.

29. Appendix B.14.9 was revised to be consistent with SP 800-132.

30. The tags for references to FIPS were modified to remove the version number. The version number is provided in Appendix C.

www.ingramcontent.com/pod-product-compliance
Lightning Source LLC
Chambersburg PA
CBHW080254180526
45167CB00006B/2524